ECOLOGICAL FOOTPRINT ASSESSMENT OF BUILDING CONSTRUCTION

Authored By

Jaime Solis-Guzman

and

Madelyn Marrero

University of Seville Ave. Reina Mercedes 4-a
41007 Seville, Spain

ECOLOGICAL FOOTPRINT ASSESSMENT OF BUILDING CONSTRUCTION

advertisements or ideas contained in the Work.

Limitation of Liability:

In no event will Bentham Science Publishers, its staff, editors and/or authors, be liable for any damages, including, without limitation, special, incidental and/or consequential damages and/or damages for lost data and/or profits arising out of (whether directly or indirectly) the use or inability to use the Work. The entire liability of Bentham Science Publishers shall be limited to the amount actually paid by you for the Work.

General:

1. Any dispute or claim arising out of or in connection with this License Agreement or the Work (including non-contractual disputes or claims) will be governed by and construed in accordance with the laws of the U.A.E. as applied in the Emirate of Dubai. Each party agrees that the courts of the Emirate of Dubai shall have exclusive jurisdiction to settle any dispute or claim arising out of or in connection with this License Agreement or the Work (including non-contractual disputes or claims).
2. Your rights under this License Agreement will automatically terminate without notice and without the need for a court order if at any point you breach any terms of this License Agreement. In no event will any delay or failure by Bentham Science Publishers in enforcing your compliance with this License Agreement constitute a waiver of any of its rights.
3. You acknowledge that you have read this License Agreement, and agree to be bound by its terms and conditions. To the extent that any other terms and conditions presented on any website of Bentham Science Publishers conflict with, or are inconsistent with, the terms and conditions set out in this License Agreement, you acknowledge that the terms and conditions set out in this License Agreement shall prevail.

Bentham Science Publishers Ltd.
Executive Suite Y - 2
PO Box 7917, Saif Zone
Sharjah, U.A.E.
Email: subscriptions@benthamscience.org

BENTHAM
SCIENCE

CONTENTS

FOREWORD

The publication is a contribution to scientific development, needed to generate standards and strategies for sustainable construction. The stringency of the contents, while its clear and informative language, makes this book a necessary publication for anyone wishing to learn or expand their knowledge on the Ecological Footprint linked to environmental aspects in the construction phase of a buildings.

The construction industry is well known for its high impact on the environment, especially during the building construction. In the present book, a methodology is defined as the first step towards the creation of an effective assessment of the Ecological Footprint (EF) of this type of activity. The EF indicator methodology has been adapted by the authors to the peculiarities of the construction sector. The procedure is based on the project budget and its bill of quantities, which is organized by means of a systematic classification of the resources into three main categories: materials, manpower and machinery. A calculation model is presented with some innovative aspects, such as including food intake and worker mobility, water consumption in the construction site and indirect costs analysis which are not normally included in the methodology of the indicator; footprints associated with cropland, pasture and fishing appear due to the inclusion of food.

The methodology and all the steps which are part of the calculation are explained, and approaches are proposed, making it easy to implement the EF assessment in any building project.

Dr. Solis-Guzman and Dr. Marrero belong to the research group ARDITEC and work in construction related problems such as construction and demolition waste management, CO_2 Footprint, Ecological Footprint and recycling of construction material. They have published over 30 articles in scientific and technical journals, and participated in over 50 conferences. They also have participated in research projects at regional and European level. Since 2009, both teach the subject Sustainable Construction in the School of Building Engineering, University of Seville, Spain.

Claudia Muñoz-Sanguinetti
Researcher of the Sustainable Architecture and Construction Group - UBB CITEC
University of Bio-Bio
Chile

PREFACE

The construction industry is well known for its high impact on the environment; however, no mechanisms for its evaluation and control have yet been made readily available, due, fundamentally, to the difficulty of defining its boundaries and the corresponding pollutant emissions. The industry needs to establish the emission sources and, if applicable, its sinks and/or mitigating factors. From this perspective, buildings are directly responsible for the generation of pollutants during their construction and operation, for water and electricity use and waste generation and also, indirectly, for emissions due to the transportation of material and occupants to and from the building site.

In this framework, the Ecological Footprint (EF) methodology is adapted to the building sector, and a model which evaluates the particular case of the construction of buildings is presented. The EF is a relatively new indicator that aims to establish results that are more intuitive of the impact on the territory of economic activities. From this point of view, the relationship of the buildings with the territory, where they are located, is defined empirically and visually, and the associated impacts are identified.

Given the difficulty in establishing a general model for the building industry, the present book focuses on the study of the construction of buildings, which constitutes the most significant impact in the territory, and, undoubtedly, the most aggressive impact from the planning point of view, since the activity is performed intensively for a relatively short period of time.

A methodology is defined as the first step towards the creation of an effective assessment of the EF of this type of construction. The procedure is based on the project budget and its bill of quantities, organized by means of a systematic construction-work breakdown structure that divides the work into three major categories: materials, manpower, and machinery. Each stream generates partial footprints (*i.e.* energy, food, mobility, construction materials, and waste). The methodology is structured in such a way that it can be adapted to any construction project that has a detailed budget and that is based on a work breakdown system.

Therefore, the present work provides a step forward in the modelling and analysis of the effects of building activity on the environment in order to identify strategies for the reduction of this impact. The book is divided into two parts: the theoretical model for calculation of the footprint of residential buildings during their construction; and the case study of the construction of an apartment building in Huelva, Spain.

The first part describes the EF indicator and its adaptation to the building construction sector. The sources of impact are the most relevant part of the analysis due to the effect of a wide variety of building elements on the calculation of the EF. The main sources of impact are grouped into direct impacts, indirect impacts, waste, and built land. The direct impacts are those taking place directly on the construction site: energy (fuel and electricity) and water. The indirect impacts refer to the construction materials (embodied energy) and manpower (food and mobility).

The second part consists of three steps. First, the general characteristics of the project under study are summarized: land type and actions to be taken on the land (detailed construction project, land subdivision, and project development). The second step explains how to apply the proposed formulation to the impact factor parameters. Finally, in the third step, the footprints associated with these impact factors are calculated: energy, water supply, food consumption, mobility, construction materials, waste, and constructed land footprints.

In an innovative approach, the evaluation is directed to professionals in the construction sector who normally deal with project budgets and well understand the work breakdown systems employed for the classification and organization of construction work units.

The methodology, applied here to a Spanish construction project, can equally be employed for construction classification systems of different countries. The aspects that are dependent on the region where the project is located are: the workforce food intake, the electric mix, the transportation systems, and the waste generation rates.

Madelyn Marrero
University of Seville
Ave. Reina Mercedes 4-a
41007 Seville
Spain

CONFLICT OF INTEREST

The author confirms that this chapter has no conflict of interest.

ACKNOWLEDGEMENT

Ministry of Innovation and Science, through the concession of the R+D+I project: Evaluation of the EF of construction in the residential sector in Spain. (EVAHLED). 2012-2014. *Ministerio de Innovación y Ciencia, por la concesión del Proyecto I+D+i: Evaluación de la huella ecológica de la edificación en el sector residencial en España (EVAHLED). 2012-2014.*

We are grateful to Dr. Rafael Llacer-Pantion and Architect Patricia Gonzalez Vallejo for the all the diagrams and drawings that help explaining the proposed model.

List of Contributors

Jaime Solis-Guzman   University of Seville, Ave. Reina Mercedes 4-a, Seville, Spain

Madelyn Marrero   University of Seville, Ave. Reina Mercedes 4-a, Seville, Spain

Patricia Gonzalez-Vallejo   University of Seville, Ave. Reina Mercedes 4-a, Seville, Spain

Rafael Llacer-Pantion   University of Seville, Ave. Reina Mercedes 4-a, Seville, Spain

2

CHAPTER 1

The Ecological Footprint Indicator

Abstract: The Ecological Footprint (EF) calculations are generally performed by following the methodology defined by Mathis Wackernagel, based on top-down analysis, on macroeconomic data that estimates the footprints at various territorial levels: Earth, continents, countries, cities, *etc*. The present chapter establishes a reference frame, also top-down, in order to define the indicators and their relevance. The indicators have been used in the calculation of the impact of humanity on the environment, among which the EF is included. According to EF methodology, all consumptions, materials, energy, and waste absorption have their corresponding productive land requirements for their production or disposal.

Keywords: Conversion factors, Ecological deficit, Ecological footprint, Productivity factors, Productive land, Sustainability, Sustainable development, Standard productive territory.

INTRODUCTION

Indicators have been used in the calculation of the impact of humanity on the environment, among which the EF is included. A first approach to the growth indicators is traced in the models developed by Jay W. Forrester [1] in the 70s. Forrester, known worldwide as the father of Systems Dynamics, is also responsible for the dynamic theory implementation in growth modelling of population, economy, or cities. Among the models generated by Forrester, the most relevant to the present work is that of global dynamics, which employs systemic dynamics in global models.

His study focuses on two parts: the first took place in the early seventies as part of a project commissioned by the Club of Rome. The Club of Rome was an organization, composed by prominent personalities, which sought to promote stable and sustainable economic growth of humanity. The project analyses the effect of population growth and human activity in a world of limited resources. From this study arises the report "Dynamics of Growth in a Finite World" [2]. The

Jaime Solis-Guzman and Madelyn Marrero

second, developed thirty years later, determines similar objectives but with more powerful tools, and introduces concepts and definitions that did not exist in the seventies, such as: EF, overreaching, sustainability, collapse, erosion cycles, *etc.* This latest report is contained in the book "The Limits to Growth 30 Years Later" [3]. The report assesses the effect of population growth and human activity in a world of limited resources, and the behavioural modes and patterns through which the human economy interacts with the carrying capacity of the planet during the present century.

Carrying capacity is a dynamic concept, constantly changing with technological progress, consumption patterns, climate, and other factors. The term designates the number of people that in the current circumstances could be sustained on the planet without causing deterioration of the Earth's overall productivity.

Similarly, the dynamic model establishes that the world's growth and its indicators, such as the EF and human wellbeing index, are part of the predictive models in order to measure the behaviour of humanity with respect to Earth.

THE EF CONCEPT

The EF indicator was introduced by Mathis Wackernagel [4], who measured the EF of humanity and compared it with the carrying capacity of the planet. According to its definition, the EF is the amount of land that would be required to provide the resources (grain, feed, firewood, fish, and (CO_2)) of humanity [5]. By comparing the EF to the amount of land available, Wackernagel concluded that human consumption of resources currently stands 50% above the global carrying capacity [6].

EF is now considered one of the most relevant indicators for the assessment of impacts on the environment, and can also be used in conjunction with other indicators, such as the carbon footprint and water footprint [7]. The strengths of the indicator include its provision of an aggregation of multiple anthropogenic pressures and its easily understood strong conservation message. On the other hand, its main weaknesses are that neither can it cover all aspects of sustainability nor all environmental concerns, and that certain underlying assumptions are controversial [7].

A different approach establishes that the footprint can be considered the sum of the farmland, urban land and the land necessary to neutralize the pollutant emissions. It can also be enunciated as an index which measures the land area required to produce the resources consumed by citizens and to absorb the waste generated by them [6].

The indicator has been used since its inception to determine impacts on differing scales: to predict the impacts generated by mankind on Planet Earth [3], for the periodic calculation of the footprint of mankind on Planet Earth [5, 6, 8], or for periodically calculating the EF of different countries [9 - 14], cities [15, 16], neighbourhoods [17, 18], productive sectors [19 - 21] and industries [22 - 27]. In the work of Nye and Rydin [28], an innovative analysis of EF per building component is proposed.

Finally, following the procedure developed by Spanish researchers [27, 29] on corporate EF calculation, the process of building construction is studied [30 - 34]. This methodology, adapted to the unique characteristics of the construction sector, has been chosen for its comprehensibility, transparency, and adaptability [29].

AN APPROACH TO THE EF INDICATOR AS A SUSTAINABILITY PARAMETER

In order to introduce a sustainability parameter, it is necessary to define its scope first. Common concepts, such as sustainable development, sustainability, and strong sustainability, are commonly used, but what do they mean? What are the differences between them?

The most significant difference is found between the concepts of sustainable development and sustainability. Sustainable development is an objective while sustainability is a process. It means that the process of achieving sustainability (ecological, economic, social, cultural, *etc.*) will lead to the attainment the objective, thereby obtaining a sustainable economy, a sustainable society, a sustainable environment, *etc.* The system, which introduces sustainability as behaviour, can achieve the proposed objective. According to Edwards [35] there exist three perspectives on sustainable development, as seen in Fig. (**1**).

Fig. (1). The three perspectives of sustainable development.

The best known definition of sustainable development is that set out in Our Common Future Report (1987), also called the Bruntland Report, which says textually that sustainable development is one "that meets the needs of the present generation without compromising the ability of future generations to meet their own needs". This definition neither proposes the actions that should be carried out, nor the direction to follow, in order to achieve a more equal world.

There are other definitions of a sustainable society, that is to say one that reaches sustainable development. From the systems theory point of view, a sustainable society is one that has informative, social and institutional mechanisms that allow it to control the positive feedback cycles which cause exponential growth of population and capital.

A sustainable system, in both material and energy terms, and according to usage rates of sources and sinks is defined by [3]:

- The usage rate of renewable resources should not exceed its regeneration rate.
- The usage rate of non-renewable resources should not exceed the development rate of sustainable renewable substitutes.

- The emission rate of pollution should not exceed the assimilative capacity of the environment.

A sustainable society does not signify zero growth. A sustainable society encourages qualitative development, not physical expansion, and uses material growth as a tool, not as a goal.

The objective is clear: to achieve sustainable development. However, it is less clear how to achieve that objective. To this end, certain indicators are employed to guide us through the process of sustainability which can maintain planet Earth as it is today.

There are two approaches to sustainability [36]:

- Weak sustainability, also known as anthropocentric sustainability. This is based on technology and substitutive processes, which optimize the extraction, production, consumption, and recycling processes.
- Strong or ecological-centred sustainability. This is defined by socio-biology, ecological and ethical-utopian humanism. This is based on the maintenance of system resilience and carrying capacity, and guides consumption towards basic needs. Policies, in this approach, have long-term objectives.

One of the most important definitions of world sustainability can be found in Donella Meadows investigations [3]. The author proposes a number of strategies to restructure our societies towards sustainability. The most relevant include: the extension of the planning horizon (policy), better indicators (that warn of the excessive consumption of resources); the shortening of response times (of the improvement actions); minimization of the use of non-renewable resources; prevention of erosion of renewable resources; the use of all resources with maximum efficiency; the deceleration and halting of the exponential growth of the population and physical capital. Through these strategies, our planet can change the present course of overreaching and collapse and can regain a sustainable growth scenario.

In order to improve indicators and reduce response times, the same allow the definition of goals. The most interesting are the synthetic indicators since they group large amounts of information and enable the application of behavioural guidelines of the various entities. The most commonly used synthetic indicators include:

- ISEW: Index of Sustainable Economic Welfare
- GPI: Genuine Progress Indicator
- SDP: Sustainable Net Domestic Product
- HDI: Human Development Index
- EF: Ecological Footprint

The synthetic indicator that better establishes the relationship between humans and their ecosystem is the EF. The main controversy that surrounds this EF indicator is whether it is a tool that measures sustainability, or whether it just visualizes human impact relative to the planet's carrying capacity. This issue arises from its basic definition: "The amount of land needed to supply resources and absorb the emissions of a given population". According to this definition, the correct analysis would be the latter, since the EF transforms, into productive land, the energy and resources required to produce and transport consumer products and the land necessary to absorb the impacts of production, transport and waste. It is therefore not a holistic approach to measure sustainability, since the EF indicator draws no dynamic images of changing conditions, and says nothing about the quality of life.

Similarly, it defines the sustainability concept as a future and equality concept. According to the definition of sustainable development, resource storage must be preserved for future generations. However, EF doesn´t note the specific resource security, although it presents the area that exceeds the reasonable consumption of resources. Neither does it take into account the use of renewable resources nor renewable energy sources. In order to measure sustainability, this indicator must complete its calculation with a sensitivity analysis of the behaviour of the variables.

DETERMINATION OF THE EF INDICATOR

According to EF methodology, all consumptions, materials, energy, and waste absorption have their corresponding productive land requirements for their production or disposal.

The productive land categories are:

• Food: agriculture, livestock, fishing
• Forestry sector
• Consumer goods
• Energy consumption (energy)
• Territory used directly (cities, infrastructure ....)

And in terms of the needs of productive land, these are (See Fig. **2**):

• CO_2 absorption
• Crops
• Pastures
• Forests
• Productive sea
• Productive land used directly

Fig. (2). Types of productive land [33].

Consumption Calculation

The total consumption of materials and energy is calculated by counting the annual consumption of food, forest products, direct annual energy consumption and that of other materials, both manufactured and natural. Consumption has three components: the materials and energy production, imports, and exports. This last component is counted negatively on total consumption.

Consumption is transformed into productive land by means of the following equation [16]:

$$AA_i = C_i/P_i \qquad \qquad \ldots \qquad \qquad (1)$$

where:

AA: is the area of productive land for each category (ha)

C: is the total consumption in metric tons (t) or gigajoules (GJ)

P: is the productivity in (t/ha) or (GJ/ha)

The results can be expressed per capita:

$$aa_i = AA_i/N \qquad \qquad \ldots \qquad \qquad (2)$$

where:

aa_i: represents the production land of each category per capita (ha/cap).

N: is the size of the population being analysed.

Finally, the EF is obtained:

$$ef_N = \Sigma aa_i \qquad \qquad \ldots \qquad \qquad (3)$$

ef is expressed in hectares per capita per year (ha/cap/ year).

Alternatively, the results can be expressed for the whole population by the equation:

$$EF = N*ef \qquad \qquad \ldots \qquad \qquad (4)$$

EF is given in hectares per year (ha/year).

Types of Productive Land

The types of productive land (according to the International Union for Conservation of Nature, IUCN), already cited in previous sections, are the following:

- Land for CO_2 absorption: This is defined as the area of forest required to absorb CO_2 emissions due to the consumption of fossil fuels for energy production. The consumption in the production of goods, housing, and transportation, among others, are included within this category.
- Agricultural land (crops): This is the area to produce the fruit, pulses, cereals and vegetables consumed. It constitutes the most ecologically productive land and generates biomass.
- Grazing land: This is the land for cattle grazing.
- Forest land: This is used for the production of forest products (wood and paper).
- Built land: This is used directly for the construction of buildings and infrastructure.
- Productive sea: This is sea which produces seafood.
- Land reserved for biodiversity: (see section "Ecological deficit").

Conversion Factors

Conversion factors are needed for the transformation of data units in order to quantify the different components of the footprint. These factors allow a comparison of the consumption of diverse geographical or production sectors, by transforming them into productive land hectares. Conversion factors can be of two types [16]:

1. Performance or productivity factor: transforms consumption data into land. Its units are kg of material production per hectare per year (kg mat/ha/year). This equivalence factor compares the productivity of each land category with respect to hypothetical land whose biological productivity is the global average of all types of productive land. That is, it relates the local land biological productivity to the global land productivity.

2. Equivalence or weighting factor: allows types of land with different productivities to be aggregated, and the EFs of countries to be compared. To this end, equivalences to each land type are applied in such a way that each hectare is normalized to the global average; in other words, it becomes global hectares (gha). The units are either expressed in terms of (gha/ha) or in per capita terms (gha/cap), thereby enabling comparisons between countries or regions to be established.

Ecological Deficit

The ecological deficit is defined as the difference between the available land (capacity) and the consumed land (EF) [16]. Carrying capacity is the available local capacity, which takes into account both the land productivity and 12% of land reserved for biodiversity conservation. Fig. (**3**) represents the inequalities that exist between the consumption of different countries, which in turn generates inequalities in the values of footprint. Developed countries have high values of EF, while the underdeveloped generally have a low EF, which, in certain cases, even designates them an ecological surplus.

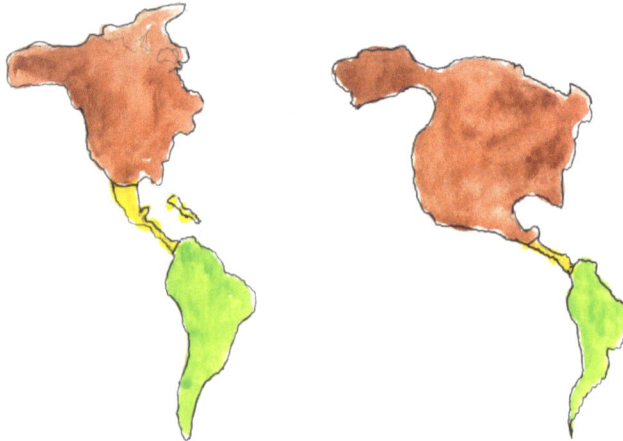

Fig. (3). Map of The Americas' consumption proportions with respect to its surface.

APPLICATION SCALES OF THE EF INDICATOR

Planetary Scale (Living Planet Model)

First, the most general case is analysed, which is the humanity footprint on Planet

Earth [5]. This requires taking into account all global consumption for all categories, as defined in the previous section, and considering all the productive land on Earth.

In order to provide the EF results, not in local productive land terms, but worldwide, the authors Rees and Wackernagel [4] devise a methodology for converting different types of areas (agriculture, pastures, forests, productive sea and used directly land) into one single area that includes them all. This would facilitate gathering each of the categories into a new concept: the standard productive territory, whose productivity factor is 1, and defines the weighted EF concept. This standard productive territory is expressed in global hectares, that is, those hectares with average biological productivity worldwide. For example, Spain´s EF is among the highest, close to 6 gha per person per year, while the global EF is about 3 gha per person per year. The planet biocapacity (carrying capacity) in 2005 was 2.1 gha per person per year, and therefore the Earth is in an overreaching state [5].

The calculations of the global and national footprints use productivity factors (Table **1**) to account for national differences in biological productivity (for example, tons of wheat per hectare produced in Britain or Argentina, compared with the global average), and equivalence factors to account for differences in average global productivity among different types of land (for example, global average forest productivity can be compared to the world average agricultural land).

Table 1. Factors of productivity for selected countries [8].

Scope	Main agricultural land	Forest	Pastures	Marine fisheries
World	1.0	1.0	1.0	1.0
Algeria	0.6	0.0	0.7	0.8
Guatemala	1.0	1.4	2.9	0.2
Hungary	1.1	2.9	1.9	1.0
Japan	1.5	1.6	2.2	1.4
Jordan	1.0	0.0	0.4	0.8
N. Zealand	2.2	2.5	2.5	0.2

(Table 1) contd.....

Scope	Main agricultural land	Forest	Pastures	Marine fisheries
Lao RPD	0.8	0.2	2.7	1.0
Zambia	0.5	0.3	1.5	1.0

The EF methodology only determines factors for agricultural land, livestock and forestry and avoids the calculation of other types of productive land that can be assimilated into the main classification. For the other territories, EF uses the following hypotheses:

1. The territory used directly is considered as agricultural land, since most of the infra-structure and built environment is located in areas of agricultural quality. This statement does not correspond to the reality of many specific situations, but is accepted as a hypothesis for the calculation.
2. The CO_2-absorption territory is considered homologous to the forest.
3. The productive sea has a productivity factor 1 because it is assumed that the sea areas are equally productive wherever they are. There are some exceptions, but this applies to most countries.

Undoubtedly, these three cases are simplifications, but are deemed necessary in all calculations in order to achieve relevant results. For example, in the case of Andalusia in Spain, available data is presented in Table **2**.

Table 2. Productivity factors of Andalusia, Spain.

	Agricultural land	Forests	Pastures	Productive sea
Acosta *et al.* [16]	1.20	0.33	1.10	1.00
Calvo [37]	1.22	0.24	1.09	1.00

Equivalence factors Table **3** are used in order to compare the global productivity of different types of territories. Each category has an intrinsic productivity that must be considered in the final calculation due not only to its ecological consideration, but also to its implications in society. Cropland represents the most valuable category, since it designates "useful" energy for humans other than pasture or forest. This hypothesis enables certain comparisons to be made, such as the number of hectares of forest that are equivalent to one of cropland.

Table 3. Equivalence factors 2003 [8].

	gha/ha
Primary cropland	2.21
Marginal cropland	1.79
Forest	1.34
Permanent pasture	0.49
Marine	0.36
Inland waters	0.36
Built-up area	2.21

The conversion factor is also needed in the calculation of the global or national EF. Conversion factors are used in the comparison of specific global hectares, and are evaluated each year in terms of the annual change in bioproductive hectares and average global productivity per hectare.

In order to be able to compare different year results of the footprint and the biocapacity, tendencies are presented in terms of constant global hectares from 2003. Similar to the use of inflation-adjusted monetary units in economy statistics, the use of constant global hectares enables any change over time to be observed in terms of absolute consumption and bioproductivity, instead of simply displaying a proportional rate between the two years. Table **4** shows the conversion of specific global hectares of some selected years into constant global hectares as of the year 2003.

In order to avoid the inclusion of exaggerated human demand on nature, the global EF considers only the aspects of resource consumption and waste production for which the Earth has a regenerative capacity and for which there is information that allows this demand to be expressed in terms of productive area. For this reason, the EF leaves aside freshwater consumption, although the energy used in the extraction, treatment and distribution of water processes is taken as part of the calculations.

The total EF of a nation or humanity is calculated based on the number of people who consume resources. These resources are taken to be the average quantity and the intensity of goods and services consumed by an average person. The footprint is an historical indicator; it does not predict how these factors change in the future. However, if the population increases or decreases (or if any other factor changes), this is reflected in the future footprint or biocapacity value. The footprint can also show how consumption is distributed among region resources. For example, the total footprint of the Asia-Pacific area, with a high population but with a low footprint per person, can be compared with that of North America, with a much lower population but with a much larger footprint per person [5].

Table 4. Conversion factors [10].

Year	2003 gha/ha
1961	0.86
1965	0.86
1970	0.89
1980	0.90
1985	0.92
1990	0.95
1995	0.97
2000	0.99
2003	1.00

The footprint determines what happened in the past; it quantifies the ecological resources used by an individual or a population, but it cannot prescribe what should be consumed in the future. Resource allocation is a policy issue, based on what society considers equitable or not; the EF is a tool which could aid resource allocation decision-making and eco-design.

The EF describes human demand on nature. There are currently 2.1 global hectares of biocapacity available per person on Earth -even less if biological productivity needed for wild species is considered [5]. The value that society assigns to biodiversity will determine the size of the buffer zones assigned to biodiversity. Efforts to increase biocapacity, such as monoculture and pesticide application, can also increase the pressure on biodiversity, which means having to increase the size of the buffer required in order to reach the same results.

The footprint analysis reflects both the increases of renewable resources productivity (for example, if the productivity of agricultural land is increased, then the footprint of a ton of wheat decreases) and the benefits of technological innovations (for example, if the paper industry doubles the general efficiency in the production of paper, the footprint per ton of paper will be halved). The EF interprets these changes and determines the degree of innovation success in helping humanity demand stay within the limits of the Planet's ecosystem capacity.

Urban Scale

For the assessment of the impacts of cities on their surrounding environment, it is necessary to previously identify the planning elements that can affect the environment. These include:

• The city size and settlement type (centralization *versus* decentralization)
• Location of housing within the city (high or low population density)
• Residential areas
• Types of housing (single-family or collective)

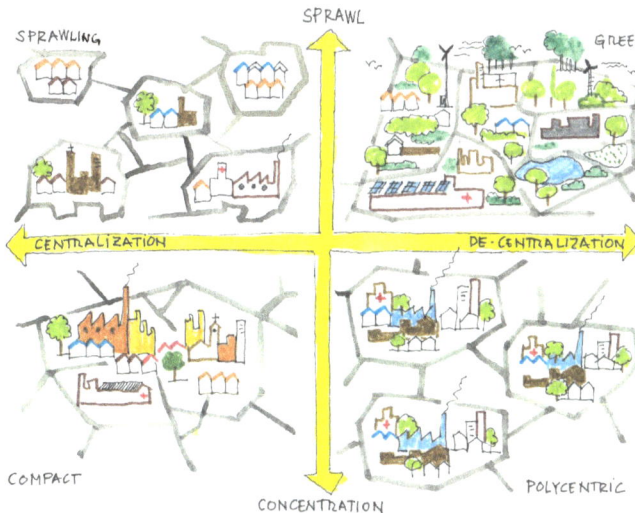

Fig. (4). The city models.

In Fig. (**4**), various city models, according to Holden [38], are represented, the ideal being the polycentric model, also called the decentralized concentration model. Within urban planning, the EF indicator assesses the impact of these different city models. Developed countries average EF is 6.5 ha/year, while in low-income countries the average is 0.8 ha/year. Moreover, the family EF has been established, according to certain studies [38], as 1.5 ha/year for high density and 2 ha/year for low density populations. From this data, a number of influencing factors can be established which decrease the EF, such as high density in residential areas, reduction in the distance to the city centre, and dense/concentrated housing design.

The National Footprint [5] has standardized calculations of the global and national EFs; however there are two principal ways to calculate the city or regional footprint.

The approach **"based on processes"**, also called **"EF analysis percomponent"**, uses formulas and supplementary statistics to allocate the national footprint per capita to consumption categories (such as food, housing, mobility, goods, and services). The averages of the per-capita footprint at the regional or municipal level are calculated by shifting these national results up or down, and clearly show the differences between national and local consumption patterns.

The **"input-output"** approach uses input tables (resources) and monetary production, physical or hybrid, in order to assign global demand to each consumption category. The I-O model developed by Wassily Leontieff [39] considers that, in order for any product to be manufactured, it is necessary to use ordinary inputs or capital goods proceeding from various productive sectors, and installations, machinery, *etc*.

The approach based on processes, for the analysis of regions or organizations, has the following EF elements [40]:

1. Electricity consumed by households
2. Gas consumed by households
3. Electricity (other uses)
4. Gas (other uses)

5. Glass recycling
6. Paper and cardboard recycling
7. Metal recycling
8. Compost recycling
9. Other household waste recycling
10. Household waste
11. Commercial waste (paper, metal, *etc.*)
12. Inert waste (bricks, cement, *etc.*)
13. Food
14. Wood products
15. Travel by car
16. Travel by bus
17. Travel by train
18. Travel by plane
19. Freight transport by road
20. Freight transport by train
21. Freight transport by shipping
22. Freight transport by plane
23. Water consumption per household

The EF methodology on a city scale uses hybrid models that mix the two approaches, although, at national level, they always remain based on the National Footprint model.

STRENGHTS AND WEAKNESSES

There are several strong aspects concerning the EF methodology, such as its simplicity, ease in calculation, and the fact that it can be understood and adopted by the general public [41]; an indicator that is easy to understand and reliable can influence consumers' decisions, legislation, and regulation [42]. The main differentiating aspects are the aggregation of factors of different sources and the importance of productivity changes.

The aggregation of factors from various sources into a single indicator normally gives only a general perspective of all impacts within an activity or productive sector [43]. On the other hand, methodologies that include several indicators can be preferable because they prevent the overlapping of impact categories [44], and the aggregation is a subjective process based on strong hypotheses in order to express all results in a single unit [45]. It is also true that the EF can be studied per category (different land classifications), which aids in the identification of the most influential aspects [23].

The footprint analysis reflects both the increases in the productivity of renewable resources (for example, if the productivity of agricultural land is increased, then the footprint of a ton of wheat decreases), and the benefits of technological innovations (for example, if the paper industry doubles the general efficiency in the production of paper, the footprint per ton of paper is halved). The EF interprets these changes and determines the degree of innovation success in helping the demands of humanity stay within the limits of the Planet's ecosystem capacity. However, the EF fails to differentiate between sustainable and non-sustainable land usage. This limitation can be considered of major importance. For example, in agriculture, intensive land exploitation can attain a higher productivity and a lower EF, but it can also deteriorate the productivity for future generations by means of soil impoverishment, pesticides, fertilizers and water consumption [46].

The strengths and weaknesses of EF can be summarized as follows [7].

Strengths:

1. Is easy to communicate and understand with a strong conservation message.
2. Provides an aggregated assessment of multiple anthropogenic pressures.
3. Allows the benchmarking of human demand for renewable resources and of carbon uptake capacity with the natural supply, and the determination of clear targets.

Weaknesses:

1. Cannot cover all aspects of sustainability, neither can it cover all environmental concerns, especially those for which no regenerative capacity exits.
2. Shows pressures that could lead to degradation of natural capital (*e.g.* reduced

quality of land or reduced biodiversity), but fails to predict this degradation.
3. Is not geographically explicit.
4. Contains a number of controversial underlying assumptions, although these have been documented.

The EF indicator has several advantages and inconveniences; many revisions have identified deficiencies and possible solutions, these are addressed in each corresponding section of the present book.

CONCLUDING REMARKS

Several indicators have been used in the calculation of the impact of humanity on the environment, among which the EF is included. The EF calculations are generally performed by following the methodology defined by Mathis Wackernagel. A reference frame is, also top-down, in order to define the indicators and their relevance. According to EF methodology, all consumptions, materials, energy, and waste absorption have their corresponding productive land requirements for their production or disposal. The most relevant aspects of the indicator are its simple concepts which are easy to calculate and can be understood by the general public, and the aggregation of factors from different sources into a single indicator. This aggregation enables the total impact of any human activity to be clearly understood.

On the down side, the aggregation is a subjective process that is normally based on strong hypotheses which allow all impacts to be represented in a single unit. However, the EF can also be studied per category (different land classifications), which helps the most influencing aspects in the global footprint to be identified.

The indicator has been used since its inception to determine impacts on differing scales: to predict the impacts generated by mankind on Planet Earth, for the periodic calculation of the footprint of mankind on Planet Earth or for periodically calculating the EF of different countries, cities, neighbourhoods, productive sectors and industries.

This methodology, adapted to the unique characteristics of the construction sector, has been chosen for its comprehensibility, transparency, and adaptability.

CONFLICT OF INTEREST

The author confirms that this chapter has no conflict of interest.

ACKNOWLEDGEMENT

Ministry of Innovation and Science, through the concession of the R+D+I project: Evaluation of the EF of construction in the residential sector in Spain. (EVAHLED). 2012-2014. *Ministerio de Innovación y Ciencia, por la concesión del Proyecto I+D+i: Evaluación de la huella ecológica de la edificación en el sector residencial en España (EVAHLED). 2012-2014.*

REFERENCES

[1]   J.W. Forrester, *Urban Dynamics.* MIT Press: Cambridge, Massachusetts, 1970.

[2]   D.H. Meadows, and D.L. Meadows, *Limits to Growth: Report to the Club of Rome on the Predicament of Mankind.* Economic Culture Fund: Mexico, 1973.

[3]   D.H. Meadows, J. Randers, and D. L. Meadows, *The Limits to Growth 30 years later.* Gutenberg Galaxy: Barcelona, Spain, 2006.

[4]   M. Wackernagel, and W. Rees, *Our Ecological Footprint: Reducing Human Impact on the Earth. British Columbia, Gabriola Island.* New Society: British Columbia, Gabriola Island, 1996.

[5]   WWF (WWF International), Global Footprint Network, ZSL (Zoological Society of London), *Living Planet Report 2008.* WWF, Gland, Switzerland, 2008. ISBN 978-2-88085-292-4. http://assets. panda.org/downloads/lpr2008.pdf [Accessed 10th August 2011].

[6]   WWF (WWF International), Global Footprint Network, ZSL (Zoological Society of London), *Living Planet Report 2010.* WWF, Gland, Switzerland, 2010. ISBN 978-2-940443-08-6. http://assets. panda.org/downloads/lpr2010.pdf [Accessed 10th August 2011].

[7]   A. Galli, T. Wiedmann, E. Ercin, D. Knoblauch, B. Ewing, and S. Giljum, "Integrating Ecological, Carbon and Water footprint into a "Footprint Family" of indicators: Definition and role in tracking human pressure on the planet", *Ecol. Indic.,* vol. 16, pp. 100-112, 2012. [http://dx.doi.org/10.1016/j.ecolind.2011.06.017]

[8]   WWF (WWF International), Global Footprint Network, ZSL (Zoological Society of London), *Living Planet Report 2006.* WWF, Gland, Switzerland, 2006. ISBN 2-88085-272-2. http://assets. panda.org/downloads/lpr2006.pdf [Accessed 10th August 2011].

[9]   A. Fricker, "The ecological footprint of New Zealand as a step towards sustainability", *Futures.,* vol. 30, pp. 559-567, 1998. [http://dx.doi.org/10.1016/S0016-3287(98)00059-7]

[10] M. Lenzen, and S.A. Murray, "A modified ecological footprint method and its application to Australia", *Ecol. Econ.,* vol. 37, pp. 229-255, 2001. [http://dx.doi.org/10.1016/S0921-8009(00)00275-5]

[11]   S. Medved, "Present and future ecological footprint of Slovenia - the influence of energy demand scenario", *Ecol. Modell.,* vol. 192, pp. 25-36, 2006.
[http://dx.doi.org/10.1016/j.ecolmodel.2005.06.007]

[12]   D.P. van Vuuren, and E.M. Smeets, "Ecological footprints of Benin, Bhutan, Costa Rica and the Netherlands", *Ecol. Econ.,* vol. 34, pp. 115-130, 2001.
[http://dx.doi.org/10.1016/S0921-8009(00)00155-5]

[13]   T. von Stokar, M. Steinemann, B. Rüegge, and J. Schmill, *Switzerland's ecological footprint: A contribution to the sustainability debate.* Federal Office for Spatial Development (ARE), Agency for Development and Cooperation (SDC), Federal Office for the Environment (FOEN), Federal Statistical Office (FSO): Neuchâtel, 2006.

[14]   M. Wackernagel, C. Monfreda, K.H. Erb, H. Haberl, and N.B. Schulz, "Ecological footprint time series of Austria, the Philippines, and South Korea for 1961–1999: comparing the conventional approach to an 'actual land area' approach", *Land Use Policy.,* vol. 21, pp. 261-269, 2004.
[http://dx.doi.org/10.1016/j.landusepol.2003.10.007]

[15]   J. Barrett, H. Vallack, A. Jones, and G. Haq, *A Material Flow Analysis and Ecological Footprint of York. Technical Report.* Stockholm Environment Institute: Stockholm, Sweden, 2002.

[16]   G. Acosta Bono, J. González Daimiel, M. Calvo Salazar, and F. Sancho Royo, *Estimación de la Huella Ecológica en Andalucía y Aplicación a la Aglomeración Urbana de Sevilla (Estimation of the Ecological Footprint in Andalusia and Application to the Urban Agglomeration of Seville).* Dirección General de Ordenación del Territorio y Urbanismo, Consejería de Obras Públicas de la Junta de Andalucía: Seville, Spain, 2001. [Online] Available: http://hdl.handle.net/10326/974 [Accessed Sept 18, 2014].

[17]   D.Z. Li, E.C. Hui, B.Y. Leung, Q.M. Li, and X. Xu, "A methodology for eco-efficiency evaluation of residential development at city level", *Build. Environ.,* vol. 45, pp. 566-573, 2010.
[http://dx.doi.org/10.1016/j.buildenv.2009.07.012]

[18]   L.W. Kuzyk, "The ecological footprint housing component: A geographic information system analysis", *Ecol. Indic.,* vol. 16, pp. 31-39, 2012.
[http://dx.doi.org/10.1016/j.ecolind.2011.03.009]

[19]   S. Gössling, C. Borgström Hansson, O. Horstmeier, and S. Saggel, "Ecological footprint analysis as a tool to assess tourism sustainability", *Ecol. Econ.,* vol. 43, pp. 199-211, 2002.
[http://dx.doi.org/10.1016/S0921-8009(02)00211-2]

[20]   C. Hunter, and J. Shaw, "The ecological footprint as a key indicator of sustainable tourism", *Tour. Manage.,* vol. 28, pp. 46-57, 2007.
[http://dx.doi.org/10.1016/j.tourman.2005.07.016]

[21]   P. Peeters, and F. Schouten, "Reducing the ecological footprint of inbound tourism and transport to Amsterdam", *J. Sustainable Tourism.,* vol. 14, pp. 157-171, 2006.
[http://dx.doi.org/10.1080/09669580508669050]

[22]   E. Holden, and K.G. Høyer, "The ecological footprints of fuels", *Transport. Res. D.,* vol. 10, pp. 395-403, 2005.
[http://dx.doi.org/10.1016/j.trd.2005.04.013]

[23]   M. Herva, A. Franco, S. Ferreiro, A. Álvarez, and E. Roca, "An approach for the application of the Ecological Footprint as environmental indicator in the textile sector", *J. Hazard. Mater.,* vol. 156, no. 1-3, pp. 478-487, 2008.
[http://dx.doi.org/10.1016/j.jhazmat.2007.12.077] [PMID: 18280032]

[24]   M. Herva, C. García-Diéguez, A. Franco-Uría, and E. Roca, "New insights on ecological footprinting as environmental indicator for production processes", *Ecol. Indic.,* vol. 16, pp. 84-90, 2012.
[http://dx.doi.org/10.1016/j.ecolind.2011.04.029]

[25]   S.D. Frey, D.J. Harrison, and E.H. Billett, "Ecological footprint analysis applied to mobile phones", *J. Ind. Ecol.,* vol. 10, no. 1–2, pp. 199-216, 2006.

[26]   V. Niccolucci, A. Galli, J. Kitzes, R.M. Pulselli, S. Borsa, and N. Marchettini, "Ecological footprint analysis applied to the production of two Italian wines", *Agric. Ecosyst. Environ.,* vol. 128, pp. 162-166, 2008.
[http://dx.doi.org/10.1016/j.agee.2008.05.015]

[27]   J.L. Domenech Quesada, *Huella ecológica y desarrollo sostenible (Ecological Footprint and Sustainable Development).* AENOR: Madrid, Spain, 2007.

[28]   M. Nye, and Y. Rydin, "The contribution of ecological footprinting to planning policy development: using REAP to evaluate housing policies for sustainable construction", *Environ. Plann. B Plann. Des.,* vol. 35, no. 2, pp. 227-247, 2008.
[http://dx.doi.org/10.1068/b3379]

[29]   J. Cagiao, B. Gómez, J.L. Domenech, S. Gutiérrez Mainar, and H. Gutiérrez Lanza, "Calculation of the corporate carbon footprint of the cement industry by the application of MC3 methodology", *Ecol. Indic.,* vol. 11, pp. 1526-1540, 2011.
[http://dx.doi.org/10.1016/j.ecolind.2011.02.013]

[30]   J. Solís-Guzmán, M. Marrero, and A. Ramirez-de-Arellano, "Methodology for determining the ecological footprint of the construction of residential buildings in Andalusia (Spain)", *Ecol. Indic.,* vol. 25, pp. 239-249, 2013.
[http://dx.doi.org/10.1016/j.ecolind.2012.10.008]

[31]   P. González-Vallejo, M. Marrero, and J. Solís-Guzmán, "Evaluation of the Ecological Footprint of residential buildings in terms of its construction typology", In: *World Sustainable Buildings Congress,* Barcelona, Spain, 2014.

[32]   P. González-Vallejo, M. Marrero, and J. Solís-Guzmán, "The ecological footprint of dwelling construction in Spain", *Ecol. Indic.,* vol. 52, pp. 75-84, 2015.
[http://dx.doi.org/10.1016/j.ecolind.2014.11.016]

[33]   M. Marrero, A. Freire-Guerrero, J. Solís-Guzmán, and C. Rivero-Camacho, "Estudio de la huella ecológica de la transformación del uso del suelo", *Seguridad y Medio Ambiente.,* vol. 136, pp. 6-14, 2014. ISSN: 1888-5438. http://www.mapfre.com/documentacion/publico/i18n/catalogo_imagenes/grupo.cmd?path=1081726. [Accessed Feb 13, 2015]

[34]   A. Freire-Guerrero, and M. Marrero, "Analysis of the ecological footprint produced by machinery in construction", In: *World Sustainable Buildings Congress,* Barcelona, Spain, 2014.

[35]   B. Edwards, *Sustainability Guide.* Gustavo Gili: Barcelona, Spain, 2008.

[36]   F. Martín-Palmero, *Desarrollo sostenible y huella ecológica (Sustainable Development and Ecological Footprint)*. Netbiblo: A Coruña, Spain, 2004.
[http://dx.doi.org/10.4272/84-84-9745-080-9]

[37]   M. Calvo, *Informe de Síntesis: Análisis Preliminar de la Huella Ecológica en España. (Synthesis Report: Preliminary Analysis of the Ecological Footprint in Spain)*. Ministerio de Medio Ambiente: Spain, 2007.

[38]   E. Holden, "Ecological footprints and sustainable urban form", *J. Housing Built Environ.,* vol. 19, pp. 91-109, 2004.
[http://dx.doi.org/10.1023/B:JOHO.0000017708.98013.cb]

[39]   W. Leontieff, *Input-Output Economics.* 2nd ed. Oxford University Press: New York, 1986.

[40]   N. Chambers, C. Simmons, and M. Wackernagel, *Sharing Nature's Interest: Ecological Footprints as an Indicator of Sustainability.* Sterling Earthscan: London, Great Britain, 2004.

[41]   B.P. Weidema, M. Thrane, P. Christensen, J. Schmidt, and S. Løkke, "Carbon footprint", *J. Ind. Ecol.,* vol. 12, pp. 3-6, 2008.
[http://dx.doi.org/10.1111/j.1530-9290.2008.00005.x]

[42]   European Commission, *Integrated Policy Product. Development of the Environmental Life Cycle Concept.* COM: Brussels, Belgium, 2003.

[43]   J. Bare, P. Hofstetter, D. Pennington, and H. Udo de Haes, "Life cycle impact assessment workshop summary. Midpoints versus endpoints: the sacrifices and benefits", *Int. J. Life Cycle Assess.,* vol. 5, no. 6, pp. 319-326, 2000.
[http://dx.doi.org/10.1007/BF02978665]

[44]   G. Finnveden, M.Z. Hauschild, T. Ekvall, J. Guinée, R. Heijungs, S. Hellweg, A. Koehler, D. Pennington, and S. Suh, "Recent developments in life cycle assessment", *J. Environ. Manage.,* vol. 91, no. 1, pp. 1-21, 2009.
[http://dx.doi.org/10.1016/j.jenvman.2009.06.018] [PMID: 19716647]

[45]   S. Giljum, E. Burger, F. Hinterberger, S. Lutter, and M. Bruckner, "A comprehensive set of resource use indicators from the micro to the macro level", *Resour. Conserv. Recy.,* vol. 55, pp. 300-308, 2011.
[http://dx.doi.org/10.1016/j.resconrec.2010.09.009]

[46]   J.C. Van den Bergh, and H. Verbruggen, "Spatial sustainability, trade and indicators: an evaluation of the ecological footprint", *Ecol. Econ.,* vol. 29, pp. 61-72, 1999.
[http://dx.doi.org/10.1016/S0921-8009(99)00032-4]

CHAPTER 2

The EF of Building Construction

Abstract: The EF indicator methodology has been adapted to the peculiarities of the construction sector during the construction phase. A calculation model is presented with some innovative aspects, such as including food intake and worker mobility, or water consumption in the construction site, which are not included in the general methodology of the indicator; footprints associated with cropland, pasture and fishing appear due to the inclusion of food.

The methodology and all the steps which are part of the calculation are explained and new hypothesis are proposed, making it easy to implement the current analysis in the EF evaluation of any dwelling construction project.

Keywords: Conversion factors, Dwelling construction, Ecological deficit, Ecological footprint, Food intake, Productivity factors, Productive land, Standard productive territory, Water consumption, Worker mobility.

INTRODUCTION

In the construction sector, the EF indicator has been applied to study the growth of high-rise districts in Tehran [1], peasant homes [2], hotels [3], and the rehabilitation of an old house [4], in order to have a developed tool to estimate the EF and carbon footprint of buildings [5]. Teng and Wu [6] analyzed the life cycle of buildings (project execution, use and demolition) and its EF (energy, resources, and solid waste CO_2), applying it to an exhibition centre in Wuhan (China). Two main studies, Bastianoni *et al.* [7] and Solís-Guzmán *et al.* [8] have chosen the EF indicator and have tried to adapt their methodology to the peculiarities of the construction sector; and both cover the construction phase of the building. Bastianoni *et al.* [7] calculated the EF of two Italian buildings, primarily considering the embodied energy of materials and the construction process (the last is estimated as 5% of the total energy of the materials). The results are reflected in land for CO_2 absorption, forest land (for wood materials), and the area occupied by buildings.

Jaime Solis-Guzman and Madelyn Marrero

Solís-Guzmán *et al.* [8] developed a similar calculation model with some innovative aspects, such as including food intake and worker mobility, or water consumption in the construction site, which are not included in the general methodology of the indicator; footprints associated with cropland, pasture and fishing appear due to the inclusion of food intake.

The last methodology has been successfully evaluated in 100 construction projects in Spain [9] and has shown itself to be sensitive to changes in building characteristics, such as whether the dwelling is detached, is of one or two floors, and whether it is a multi-family dwelling. These changes significantly affect the results. The methodology and structure defined can be adapted for use in other countries by means of regional or national construction breakdown systems, by following the proposed steps: define the budget quantities into construction systems; then transform these into impacts such as materials, manpower and machinery; and finally, determine the resources consumed and their corresponding EF.

In the present book, the previous model has been improved in order to predict the power demand on a construction site and the EF of the worker's meal. The indirect costs of the project budget that cover the rental of temporary offices and toilet facilities, and the leasing of cranes, among other general expenses, are analysed from an electricity-intensity standpoint; this power consumption can be also established for various projects.

Solis-Guzman *et al.* [8] and Gonzalez-Vallejo *et al.* [9], based on Domenech's work [10], calculate the EF of the worker's meal from an economical point of view, where a 10-Euro cafeteria menu is evaluated. A new methodology is proposed where the reports from the Food and Agriculture Organization of the United Nations (FAO) [11] and the Global Footprint Network (GFN) [12] are analysed and combined (weighted averages) in order to determine the footprint of a daily meal.

As stated in Chapter 1, there are two well-recognized methodologies to measure EF: the component method and the compound method. While the compound method is based on national statistics of input-output flows (production, import, export), the component method is based on life-cycle data for each individual component involved in calculations. In general, in the case of the application of EF to any type of product, the application of the component method is preferred for the EF assessment, since it is based on the real life-cycle data of the processes. Furthermore, either of two perspectives, namely additive and mutually exclusive use of land, can be adopted.

In the present methodology for building construction, the component method is applied to materials (life-cycle data), and to direct machinery consumptions. However, input-output flows are also used in the food consumption analysis, where international statistics are employed in order to determine the average worker's meal, the food productivity, and energy intensity.

In the present book, the last methodology and all the necessary calculations are explained in detail, making easy to implement them in the EF evaluation of any dwelling construction project.

OBJECTIVES

The objectives focus on developing a methodology to calculate the EF of residential buildings. The area of study is centred on the construction, in the first phase of the building process, as shown in Fig. (**5**).

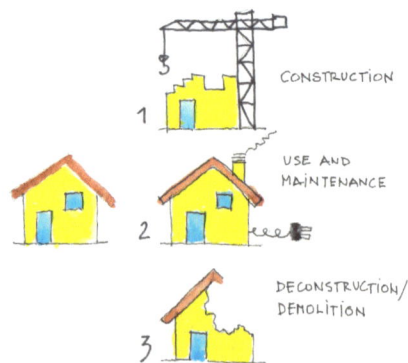

Fig. (5). Dwelling life cycle: dwelling construction, usage and maintenance, and demolition/deconstruction.

The complexity of the EF calculations limits the ability to extend the investigation to the other two phases of the life cycle of buildings, those of use and maintenance, and of demolition/deconstruction. Therefore, the analysis of the footprint covering the whole life of the building (construction, use, maintenance and demolition) is not included in the present methodology.

METHODOLOGY

The methodology, for the calculation of the EF of the residential sector during the construction phase, defines the impact sources of the EF, as they represent the generators of consumption in the territory (Fig. **6**). These sources lie in the top level of the flowchart of the methodology:

- Direct consumptions
- Indirect consumptions
- Waste generation
- Constructed area

Direct consumption is that which is directly used on the construction site, through energy expenditure (in the form of fuel or electricity) or expenditure of water. Both are located in the second level of the flowchart, and are considered as resources (see Fig. **6**). Indirect consumption is caused by the indirect usage of resources, such as material or energy resources from other previously used resources, which are:

- Manpower
- Building materials

Manpower consumption in building construction involves food expenditure by the operators, and also fuel involved in their transportation (commuting to the construction site).

Building materials, during their manufacturing, transportation and installation processes (see Fig. **6**), consume fuel (transport of materials to the workplace) or energy (required in the manufacturing of materials and installation). In EF analysis, the quantitative study of building materials is not relevant, but how the quantities consumed are translated into resources that are expressible in terms of the EF. Therefore this quantity is transformed into terms of primary energy consumption, as performed with electricity and mobility.

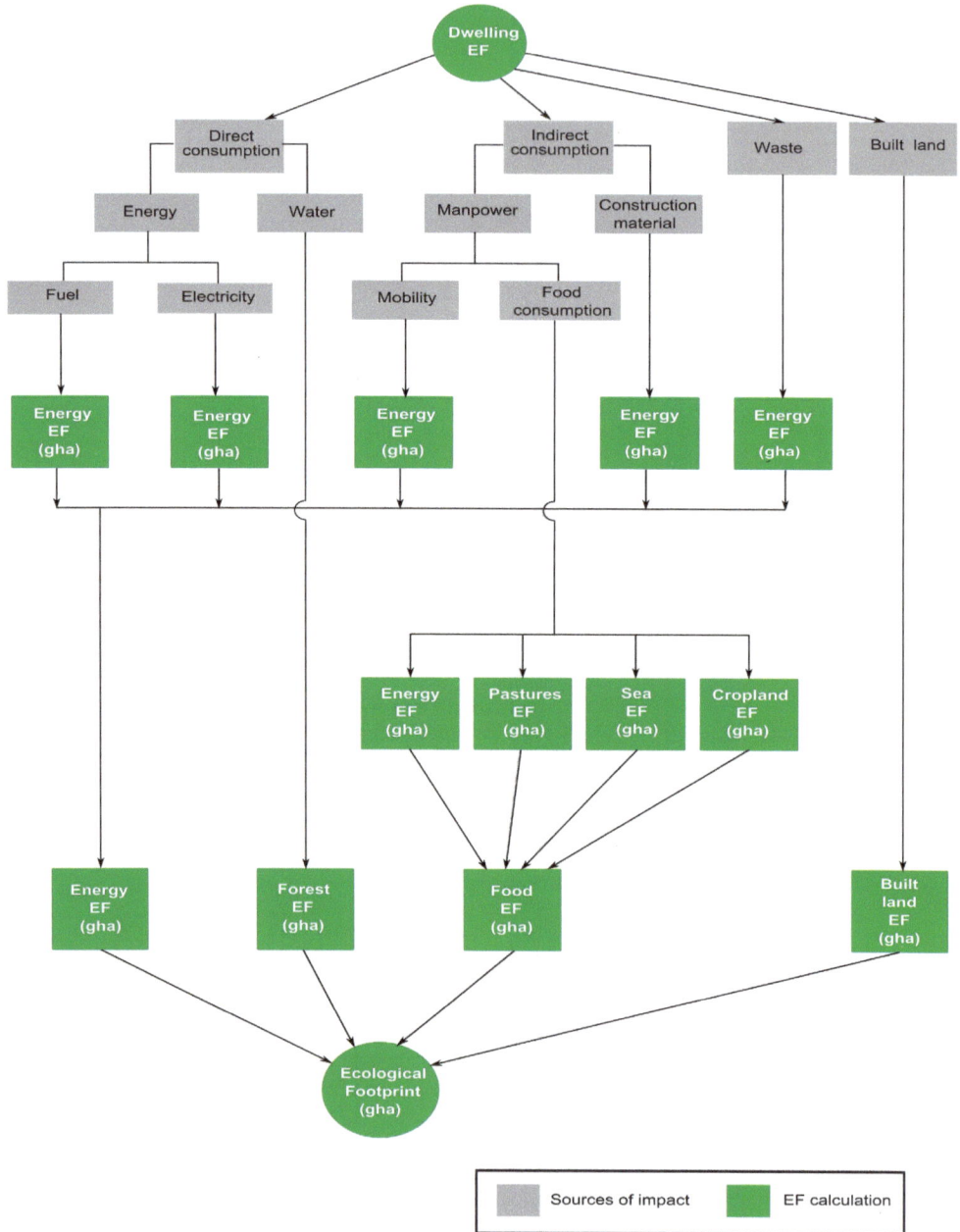

Fig. (6). The methodology flowchart.

The third impact factor is created by the waste generated on the construction site: primarily, construction and demolition waste (CDW). The final impact source to be analysed is the constructed area, which causes land consumption, and, therefore, a footprint. Each of the impact factors uses resources (energy, water, manpower, materials) or generates waste.

Intermediate elements (Fig. **6**) transform the previously described consumption into elements that allow the definition of the various footprints that make up the global footprint of the system under study. These elements are given in detail in the following sections.

- The intermediate elements include:
- Fuel
- Electricity
- Mobility
- Transportation of materials to the worksite
- The manufacture of building materials
- CO_2 emissions
- The land needed to absorb CO_2 emissions
- The territory occupied by the various impact sources

The various coefficients that enable the transformation of the consumption and intermediate elements into partial footprints are:

- Electric mix
- Forest productivity
- Food performance factor
- Mobility coefficient
- Transport coefficient
- Embodied energy coefficient
- Waste generation coefficient
- Waste conversion factor
- Consumed land
- Emission factor
- Absorption factor

- Equivalence factors
- Productivity factors

Each of these factors is explained in the following corresponding section depending on its impact source.

The various partial and total EFs generated in dwelling construction are obtained from the intermediate elements and their corresponding coefficients. They are located in the penultimate level of Fig. (**6**), and are represented by green squares:

- Forest
- Food
- Energy
- Land occupied directly

Once the theoretical part is defined, then a real case can be analysed in the practical part. This requires the identification of the activity boundaries and the scale on which they work. The working scale is the urbanization, and for the boundary, the activity is limited to transforming the land from rustic (rural) to urban. Furthermore, building typologies are then defined, in order to apply the methodology developed in the theoretical part.

The practical part has the following sequence:

1. Analysis and evaluation of material and energy consumption during the construction phase of the building typology considered (building system footprint). This evaluation includes: consumption of all material resources used in building construction, energy consumed in the construction site (including earthworks), worker mobility and energy consumed during the transportation and treatment of waste generated on the construction site.
2. Analysis and evaluation of the consumption of material and energy resources in the construction phase of the urbanization typology considered. This step includes the same aspects as the previous step. This analysis is performed to take into account all the impacts generated by the building project, including the urbanization where the building under study is located.
3. Analysis and evaluation of energy consumption linked to the water supply used

in the construction.

4. Analysis and evaluation of the consumption of material and energy resources by the manpower employed in the construction.
5. Calculation of the partial footprints.
6. Calculation of the EF produced by the construction and urbanization of the building typology studied.

CONCLUDING REMARKS

Footprint studies are primarily focused on an urban scale, thereby making it difficult to extrapolate information to the scale of individual buildings. The definition of the measurement units of the indicator for buildings is complicated due to the peculiarities of construction activity. This is overcome, in the proposed methodology, by means of a bottom-top analysis which starts at the project bill of quantities.

The EF indicator methodology has been adapted to the peculiarities of the construction sector during the construction phase. A calculation model is presented with certain innovative aspects, such as the inclusion of food intake and worker mobility, and water consumption on the construction site, which are not included in the general methodology of the indicator; footprints associated with cropland, pasture and fishing therefore appear due to the inclusion of food. The methodology and all the steps which are part of the calculation make the current analysis easy to implement in the EF evaluation of any construction project.

The analysis also depends on the periodic review of statistical data that defines waste generation, electric mix, electricity productivity, the nutritional composition of meals, construction material embodied energy, *etc.*

The difficulty of establishing the overall costs of a project as adjusted to a standard work breakdown system is evident since most construction companies often have their own cost databases. Furthermore, for the calculation of the overall costs, it is necessary to determine the direct and indirect costs in full, with the subsequent difficulty of integrating these costs into the methodology of calculation of the indicator. The EF has always avoided the calculation of indirect costs associated with any business. This can be overcome by grouping all work units or activities that take place on the construction site into three major categories: manpower (food and mobility), machinery (fuel and electricity), and/or construction materials (embodied energy).

In an innovative approach, the present evaluation is directed towards those professionals in the construction sector who normally deal with project budgets and well understand the work breakdown systems employed for the classification and organization of project work units. This division allows a general approach to the EF assessment of construction projects.

CONFLICT OF INTEREST

The author confirms that this chapter has no conflict of interest.

ACKNOWLEDGEMENT

Ministry of Innovation and Science, through the concession of the R+D+I project: Evaluation of the EF of construction in the residential sector in Spain. (EVAHLED). 2012-2014. *Ministerio de Innovación y Ciencia, por la concesión del Proyecto I+D+i:Evaluación de la huella ecológica de la edificación en el sector residencial en España (EVAHLED). 2012-2014.*

REFERENCES

[1]   P. Samadpour, and Sh. Faryadi, "Determination of ecological footprints of dense and high-rise districts, case study of Elahie neighbourhood, Tehran", *J. Environ. Studies,* vol. 34, no. 45, pp. 63-72, 2008.

[2]   X.Y. Zhao, and X.W. Mao, "Comparison environmental impact of the peasant household in han, zang and hui nationality region: Case of zhangye, Gannan and Linxia in Gansu Province, Shengtai Xuebao", *Acta Ecol. Sin.,* vol. 33, no. 17, pp. 5397-5406, 2013.
[http://dx.doi.org/10.5846/stxb201206050813]

[3]   B. Li, and D.J. Cheng, "Hotel ecological footprint model: Its construction and application", *Chinese J. Ecol.,* vol. 29, no. 7, pp. 1463-1468, 2010.

[4]   G. Bin, and P. Parker, "Measuring buildings for sustainability: Comparing the initial and retrofit ecological footprint of a century home – The REEP House", *Appl. Energy,* vol. 93, pp. 24-32, 2012.
[http://dx.doi.org/10.1016/j.apenergy.2011.05.055]

[5]   V. Olgyay, "Greenfoot: A tool for estimating the carbon and ecological footprint of buildings", In: R. Campbell-Howe, Ed., *American Solar Energy Society - SOLAR 2008, Including Proc. of 37th ASES Annual Conf., 33rd National Passive Solar Conf., 3rd Renewable Energy Policy and Marketing Conf.: Catch the Clean Energy Wave.* 2008, pp. 5058-5062.

[6]   J. Teng, and X. Wu, "Eco-footprint-based life-cycle eco-efficiency assessment of building projects", *Ecol. Indic.,* vol. 39, pp. 160-168, 2014.
       [http://dx.doi.org/10.1016/j.ecolind.2013.12.018]

[7]   S. Bastianoni, A. Galli, R.M. Pulselli, and V. Niccolucci, "Environmental and economic evaluation of natural capital appropriation through building construction: practical case study in the Italian context", *Ambio.,* vol. 36, no. 7, pp. 559-565, 2007.
       [http://dx.doi.org/10.1579/0044-7447(2007)36[559:EAEEON]2.0.CO;2] [PMID: 18074892]

[8]   J. Solís-Guzmán, M. Marrero, and A. Ramírez-de-Arellano, "Methodology for determining the ecological footprint of the construction of residential buildings in Andalusia (Spain)", *Ecol. Indic.,* vol. 25, pp. 239-249, 2013.
       [http://dx.doi.org/10.1016/j.ecolind.2012.10.008]

[9]   P. González-Vallejo, M. Marrero, and J. Solís-Guzmán, "The ecological footprint of dwelling construction in Spain", *Ecol. Indic.,* vol. 52, pp. 75-84, 2015.
       [http://dx.doi.org/10.1016/j.ecolind.2014.11.016]

[10]   J.L. Domenech Quesada, *Huella ecológica y desarrollo sostenible (Ecological Footprint and Sustainable Development).* AENOR: Madrid, Spain, 2007.

[11]   FAOSTAT, *Statistic division of the Food and Agriculture Organization of the United Nations,* 2014. http://faostat3.fao.org/home/E [Accessed 21st December 2014].

[12]   Global Footprint Network, *National Footprint Accounts Workbook Learning License,* 2014. http://www.footprintnetwork.org/en/index.php/GFN/page/national_footprint_accounts_license _academic_edition/ [Accessed 21st December 2014].

CHAPTER 3

Direct Consumption: Energy and Water

Abstract: The impact sources of energy and water, which consume resources directly, are analysed. Both are crucial in the EF calculation. First, for energy consumption, both fuel and electricity are examined. The transformation of these two types of consumption into EF values is performed through the existing EF methodology, although certain procedures have to be adapted to the building sector. The conversion of energy to productive territory considers forest land as the productive land necessary for the absorption of CO_2 emissions resulting from burning fuel. In the energy footprint of the building, the average absorption factor obtained from urban vegetation is applied. Using the absorption and emission factors established, the energy productivity is obtained.

Secondly, the water supply EF is evaluated. Generally, all EF studies obviate this aspect due to the intrinsic difficulty of transforming water consumption data into a quantity of consumed land; a transformation is proposed. In the water EF, the forest productivity is employed, which is taken as 1,500 m^3/ha/year.

Keywords: Absorption factor, Conversion factors, Dwelling construction, Ecological footprint, Electricity, Emission factor, Energy, Forest productivity, Fuel, Fuel productivity factor, Productive land, Water consumption.

INTRODUCTION

This section analyses the impact sources of energy and water, which consume resources directly. Both are crucial in the EF calculation. First, for energy consumption, both fuel and electricity are examined. The transformation of these two types of consumption into EF values is performed through the existing EF methodology, although certain procedures have to be adapted to the building sector. Secondly, the water supply EF is evaluated. Generally, all EF studies obviate this aspect due to the intrinsic difficulty of transforming water consumption data into a quantity of consumed land. In this section, a transformation is proposed.

ENERGY

The energy analysis follows the sequence described in Fig. (**7**), where energy consumption is grouped into two main types of consumption: fuel and electricity. Fuel is consumed by construction machines; and electricity is needed for electric machines and temporary offices and meeting rooms on the construction site, *etc.*

Electricity and Fuel

From the point of view of consumption impacts occurring in the building construction, electricity is one of the most prevalent. It is necessary to start from a primary energy source, which, throughout the twentieth century, was basically of fossil origin (coal, oil or natural gas), used in thermal plants which produced electrical energy. Alternative sources include nuclear, hydroelectric and, in the last 30 years, renewable energy plants.

It is necessary to ascertain the electrical energy sources in order to evaluate how the electricity consumption translates into the EF. In Table **5**, the energy sources in Spain for power generation are represented, and data from various bibliographic sources is provided in order to compare values.

Table 5. Electric mix in Spain.

Generation/ Percentage	Andalusia 1996 [1]	Spain 2000 [2]	Spain 2005 [3]	Spain 2006 [4]	Spain 2007 [5]
Fossil fuels	50	49.1	57.7	59.1	62.2
Coal	33	32.3	29.1	22.4	23.9
Oil	9	9.2	8.7	7.9	6.8
Natural gas	8	7.6	19.9	28.8	31.5
Hydroelectric	16	16.3	12.1	9.8	9.7
Nuclear	30	30.1	23.1	18.8	17.7
Renewable	4	4.5	7.1	10.7	10.4
Total	100	100	100	100	100

The first column shows data from Andalusia [1] and the following columns show Spain's electricity generation sources (sequentially: [2 - 5]), all with very similar values from one year to the next. A tendency to reduce the use of nuclear energy

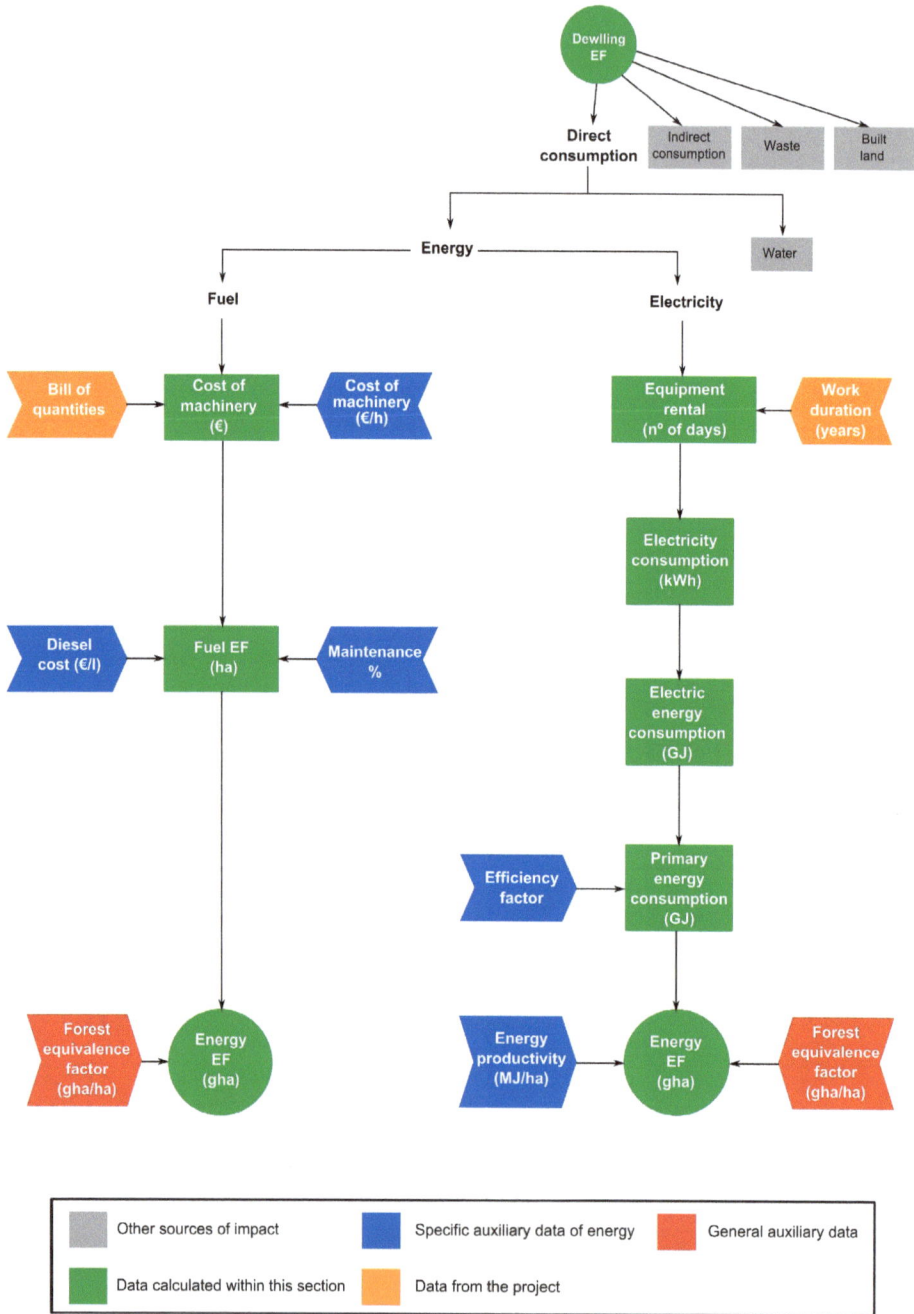

Fig. (7). Calculation of the energy footprint sequential.

and a significant increase in natural gas can be observed in the data set. The increase in renewable energy remains insignificant. The tendency over the years is the same, the percentage of the combined cycle plants is increasing (gas) and the coal and oil plants are decreasing. The percentage of nuclear energy is falling, and those of hydroelectric and renewable energy are slightly increasing (together these latter two account for 20%).

Secondly, the power plants involved in the electricity generation from primary energy sources are identified. The majority of plants are gas combined cycle plants and 1 kWh of electricity emits 0.545 kg CO_2 [6]. Yields from thermal power plants take into account the transformation of the primary sources into electric energy in the plant, but fail to consider losses that result from energy transportation, processing and distribution, which can reach approximately 10% [7], see Table **6**.

Table 6. Power plant yields.

Generation/Year	IDAE [8]	IDAE [3]
Coal	35.5	36.1
Oil	35.5	-
Natural gas combined cycle	51.0	54.0

In 2001, the performance of nuclear and thermal plants was close to 30% of the total electricity produced in Spain [1] and, in a similar way, IDAE [9] considers that the relationship between the energy source used for the generation of electricity and the electric energy obtained is:

Primary energy source = 3 * Electric energy

This implies that if the primary energy source is 1, then the electric energy is 1/3, which means a yield of 33%, similar to the 2001 estimate.

The total primary energy source consumed to produce electricity can be expressed as:

$$C = \frac{Pee}{Fef} \qquad \ldots \qquad (5)$$

where:

C: energy consumption (GJ)

Pee: production of electric energy (GJ)

Fef: efficiency factor

The efficiency factors for the various types of power plant types are shown Table **6**.

Oil is used for electric power and transportation, while natural gas, coal and nuclear energy are used only to generate electricity. The electricity consumed in construction work is considered to be obtained from thermal plants, which currently are in the majority, and its performance is taken as 30%.

The fuel consumption in Spain in 2006, at the primary energy level [9], is distributed as listed in Tables **7** and **8**. The final energy consumption by sector in 2004 is summarized in Table **9** [9]. Fuel consumption is mostly due to transportation and industrial sectors. The fuel usage in construction is intensive, because the transport of workers and freight must be added to the machinery consumption throughout the construction process. It is therefore necessary to ascertain the resulting total fuel consumption in order to calculate its corresponding EF, in the same way as electricity consumption is analysed in the previous section. Fuel processes are efficient, and their yields in machinery and transportation are close to 1 [10].

Table 7. Primary energy consumption.

Generation type	Percentage
Fossil fuels	83
Coal	13
Oil	49
Natural gas	21
Nuclear	10.50
Renewable	6.50
Total	100

Table 8. Renewable energy.

Generation type	Percentage
Wind	1.3
Hydraulic	1.6
Biomass	2.9
Biogas	0.2
Photovoltaic solar	0.03
Solar thermal	0.05
Geothermal	0.01
Biofuels	0.41
Total	6.50

Table 9. Final energy consumption per sector.

Sectors	Percentage
Transport	39
Industry	31
Home	17
Services	10
Agriculture	3
Total	100

Energy Productivity

Following the analysis of Fig. (**7**), the first step is to determine the electricity generation sources and fuel type of the energy used on the construction site. With this data, the primary energy consumption is calculated. To express this in terms of land consumption, the term *energy productivity* is employed, which is defined as the amount of land (in hectares) required to produce 1 GJ of energy.

Since the energy origin is different for the particular characteristics of energy sources, the land productivity therefore also differs. The land used for energy production has units of GJ/ha of productive land, and is hence consistent with the EF specific formulation.

The following hypotheses are used in the calculation of non-fossil energy source productivity [1]:

- Hydroelectric plant: This is the space occupied by conductors and processing instal-lations. The productivity of this type of land is 1,000 GJ/ha.
- Nuclear energy: As this is generally not considered to be an energy option, the EF methodology replaces it with another energy source, such as fossil fuels. In standard EF methodology, nuclear energy is considered as fossil origin, and identical productivity factors to those used with oil are applied. However, from the point of view of mass flow, nuclear energy emits no CO_2, thereby contradicting this hypothesis; however, it remains generally accepted.
- Renewable energies: Their productivity depends on the type of energy used. Thus, biomass is included as forest land, which is the waste produced by forests or cropland. Solar thermal energy is included within the category of constructed land, although some sources estimate its productivity in the range of 10,000 [1] to 40,000 GJ/ha [11]. Photovoltaic and wind energies have changing values, mainly due to the technology employed, which, in turn, causes yield changes. According to sources consulted, photovoltaic energy can achieve productivities of 1,000 GJ/ha, and, with technological improvements, up to 1,500 GJ/ha (this surface can be counted as built land), and wind productivity is 18,750 GJ/ha, and, with technological improvements, has reached 110,000 (60,000 ±50.000 GJ/ha) [11].

Table **10** summarises the energy productivities [12]. The analysis carried out by Wackernagel and Rees in 1996 finds alternatives for the purpose of the minimisation of fossil-fuel emission by absorption or by the use of alternative fuels, such as biomass and bioethanol. Generally, biomass is used, because the other option is rarely viable. Furthermore, their productivity values are highly variable, especially in the case of renewable energies. These values are not used in the present methodology because energy in construction comes from fossil fuels. In the cases where it is necessary to use alternative energy sources in the analysis, an average value of energy productivity is applied.

The energy sources from fossil fuels generate CO_2, which is one of the greenhouse gases (GHG), and perhaps the main agent of climate change. Therefore, the study of fossil fuels involves the analysis of the productivity of this energy type, and a similar analysis is carried out to that performed for other sources, and of the associated emission factors.

Table 10. Final energy consumption per sector.

Energy sources	Productivity (GJ/ha)
Fossils	
Replacement by biofuel (ethanol)	80
CO_2 absorption	100
Replacement by biomass	80
Hydroelectric	
Hydroelectric (average)	100
Low level	150-500
High altitude	15,000
Renewable	
Solar thermal	>40,000
Photovoltaic solar	1,000
Wind	12,500

The conversion of fossil fuels to productive territory considers forest land as the productive land necessary for the absorption of CO_2 emissions resulting from burning fuel. Burning fossil fuels destroys natural capital accumulated over millions of years, and its current use is far from sustainable.

In order to analyse fossil fuel from a sustainable approach, the consumption rate of natural capital must be equal to its natural regeneration. However, the regeneration rate of fossil fuels is extremely slow, in terms of millions of years. The only option is to invest in the creation of an equivalent capital, in this case, in planting forests.

The data on fossil fuel consumption is transformed into productive land values by taking two factors into account: CO_2 emissions per gigajoule of energy produced (emission factor); and the hectares of forest needed to absorb CO_2, also per unit of energy produced (absorption factor).

Both factors are related by the expression:

$$EP = \frac{A}{E} \qquad \ldots \qquad (6)$$

where:

EP: fuel productivity (GJ/ha)

A: absorption factor (kg CO_2/ha)

E: emission factor (kg CO_2/GJ)

The absorption factor defines the ability of the forest to assimilate CO_2, therefore it does not depend on the fuel characteristics; however, the emission factor does. In fact, a higher emission factor decreases fuel productivity. Different types of fossil fuels have different emission factors, which imply that the corresponding productivity is also different for each fossil fuel, because this productivity depends on the hectares of forest required to absorb the CO_2 released. The more CO_2 released per energy produced, the less productivity the fossil fuel has, because more hectares of forest are needed to absorb these emissions.

The absorption factor of forests and the ratio of energy per hectare vary according to whether the fuel source is coal, oil, gas, or wood, in accordance with Rees and Wacker-nagel´s methodology [11]. Initially these authors estimated an absorption factor of 1.8 tons of carbon (C) per hectare per year, and a forest maturation time of 50 to 80 years. The Intergovernmental Panel on Climate Change (IPCC) estimates forest productivity: forest maturation (crop cycle) of 40 years has an average absorption of 1.42 t C/ha per year (world average absorption) or 5.21 t CO_2/ha per year. In order to obtain these factors in CO_2 units, the formula from carbon combustion (C) and its molecular weight ratio of carbon (12) and CO_2 (44) are employed:

$$C + 2O \rightarrow CO_2 \qquad \ldots \qquad (7)$$

$$\frac{1.42\ t\ C * 44\ g\ CO_2}{12\ g\ C} = 5.21\ t\ CO_2 \quad \ldots \qquad (8)$$

This forests absorption factor is on the low side; other studies on eucalyptus yield an absorption rate of 25 t CO_2/ha per year. And the analysis conducted in Patagonia with eucalyptus plantations show absorption rates of up to 43 t CO_2/ha per year [13].

In Spain, forests are mainly formed by two types of trees: pines and eucalyptus, used for paper production. The average fixation rates for pine species are about 25 t CO_2/ha per year. For typical pastureland species, the most relevant in Andalusia being oak trees and cork oak trees, fixation rates are slightly higher than those of pine trees, between 30-40 t CO_2/ha per year.

Forests are considered, as seen above, as sinks for CO_2. However, they are not permanent sinks. This is only true while trees are growing. Mature trees have zero CO_2 net balance, because a mature forest does not detract from the atmosphere CO_2.

In the building sector, urban ecosystems, such as urban vegetation, minimize the footprint of the building. If, instead of using forest absorption factors, urban vegetation factors are used, in the form of new vegetation in the urbanizations built, then the energy footprint can vary positively or negatively.

Urban ecosystems currently cover 4% of the Earth's surface, (Fig. **8**). Vegetation is one of the elements of urban ecosystems that can be altered in order to obtain benefits. An important part of the urban ecosystem is in the form of hedgerows and woodland along the sides of roads, which has great value to city settlers.

Fig. (8). Urban ecosystems cover 4% of the Earth's surface.

The correct location of trees acts with a dual effect on the CO_2 emission rate: the sequestration rate of the trees itself and a reduction in energy consumption since long-term urban trees act as a reliever agent against the problem of CO_2 emission into the atmosphere, (see Fig. **9**). Some tree species absorb more CO_2 according to the climate. Thus, it has been shown in previous research [13] that large and vigorous trees absorb about 90 times more carbon annually than small trees (93 kg C/year compared to 1 kg C/year). These studies have been performed in forests and large parks but do not take city roadside trees into account. The CO_2-absorption capacity of urban vegetation is from roadside trees and vegetation of parks and forests located in the suburbs. Similar studies [13] have been carried out in Spanish cities in order to determine the CO_2-sequestering capacity of urban trees, and of roadside bushes and grass, the results of which are shown in Table **11**.

Table 11. Forest absorption factors [13].

Vegetation type	Absorption (kg CO_2/ha per year)
Urban woodland	
Acacia	802
Poplar	498
Judas tree	19
Brachychiton	957
Catalpa	11
Cypress	385
Japanese plum	17
Jacaranda	1,832
Laurel	384
Melia	5,969
Orange tree	555
Elm	762
London plane	478
Urban shrubs	
Oleander	31
Privet (Ligustrum)	1.3

(Table 11) contd.....

Vegetation type	Absorption (kg CO_2/ha per year)
Laurustinus	46
Lantana	6
Lentisco	0.2
Arbustus	28
Sage	0.6
Palm tree	40
Urban herbaceous	
Grass	1.5
Other trees	
Cork	4,537
Holly tree	48,870
Oak tree	5,040
Olive tree	570
Pine tree	27,180

Fig. (9). The correct location of the trees acts on the CO_2 sequestration rate and on the reduction in energy consumption [13].

For example, the absorption factor of the pine tree is very similar to those previously discussed. There are also some differences, as in the oak tree or cork oak, which has fixation rates of 30 to 40 t CO_2/ha per year while, in Table **11**, values of only around 5 t CO_2/ha per year are reached. However, the relevant data for the present methodology refers to urban trees, bushes and grass, which are found in gardens of residential buildings.

The different absorption factors of urban trees are described in the text as an example of different capacities per tree but in the present methodology a global absorption factor is used, corresponding to that of Wackernagel and Rees [12]. The absorption factor is taken as A= 5.21 tonnes of CO_2/ha per year or 1.42 tonnes of C /ha per year [11].

Once the absorption (A) and emission factor (E) are established, the next step is to define the productivity of fossil energy sources:

EP: Energy productivity of fuel (GJ/ha/year), which is considered as oil (Table **12**).

Table 12. Emission factors (E) and EP of energy sources.

Energy sources	E (kg C/GJ) [1]	EP (GJ/ha/year)
Fossil fuels		
Coal	26.00	55
Oil	20.00	71
Natural gas	15.30	93
Nuclear	20.00	71

Energy productivity of fuel is expressed as:

For liquid fuels (oil) the emission factor is 20 kg C/GJ [1]; the absorption factor is 1.42 t C/ha per year; and the energy productivity is:

$$\frac{1.42}{0.020} = 71 \text{ GJ/ha per year} \quad \ldots \quad (9)$$

That is, 1 hectare of forest can annually absorb the CO_2 emissions generated by 71 GJ of liquid fuel burned. From the equation, if the absorption rate varies, the energy productivity also varies [1]. Only fossil fuels and nuclear energy emit CO_2, other energy sources do not emit greenhouse-effect gases (GHG) into the atmosphere. Table **12** summarizes the energy productivities of different sources.

The CO_2 emission factors need to be revised periodically, as described in the GHG Inventory Report 1990-2004 (May 2006) [14].

Calculating the Footprint Associated with Energy Consumption

Before explaining the calculation procedure, the EF calculated for the energy consumption in Great Britain [15] is presented in Table **13**. The most significant footprints are those of fossil fuels, while those of renewable energy are relatively small (in the case of wind) or its footprint only comes from the construction of infrastructures. However, the data on renewable energy is evolving rapidly due to the development in cutting-edge technologies. In the previous section, the productivities of renewable sources exceed 1.000 GJ/ha [11]. If energy productivity is very high, that implies that the emission factor tends towards zero and the footprint becomes insignificant; CO_2 emissions from renewable energy (including hydroelectric) are considered irrelevant [3]. It is also important to take into account that electricity generated from fossil fuels and nuclear power stations normally covers 80% of the total (Table **7**), and the sum of renewable and hydroelectric remains at around 20%.

Table 13. EF power generation in Britain.

Power source	EF (gha/year/GWh)	Specific characteristics
Coal	198	
Oil	150	
Natural gas	94	
Wind	6	
Photovoltaic	24	panel manufactured w. fossil fuels
Biomass	27-46	Embodied energy is fossil fuels
Hydroelectric	10-75	Embodied energy is fossil fuels

From the previous data analysed, the following hypotheses for the calculation of the electricity footprint are defined:

• Only fossil fuels and nuclear sources generate an EF.
• The electricity mix is obtained, for Spain, from the Ministry of Industry [5],

collected in Table **5**, and from the electricity supply company (ENDESA is the power supply company in Seville).

• Energy productivity data is summarized in Table **12**.
• Performance data is in Table **6**.
• The absorption factors are those listed in Table **11**.

Once the hypotheses are set, the procedure is:

1. Transformation of calculated consumption (kWh) to GJ, using the conversion factor.
2. Application of the defined yields (30% efficiency):

$$0.0036/0.3 = 0.0120 \; GJ/kWh \qquad \ldots \qquad (10)$$

3. EF calculation for each of the energy sources considered, by the expression:

$$EF_{we} = \sum_i \frac{P_i}{EP_i} \cdot e_f \qquad \ldots \qquad (11)$$

where:
EF_{we}: EF of electricity consumption
P: Primary energy consumption (GJ)
EP: Energy productivity (GJ / ha)

4. Application of equivalency factors for the land production considered, in this case the forest land productivity is used Table **3**, and global hectares (gha) are obtained from the hectares (ha).

In the calculation of the EF generated by fuel consumption, there are relevant previous results, such as in the British footprint, whereby the natural gas EF is between 45 and 49 gha per year per GWh ([15] and [16], respectively), and the oil EF is between 59 and 62, ([15] and [16], respectively).

In this section, only the EF of fuel used in the building construction is considered; the mobility footprint is analysed in a different section. Wackernagel and his team add 50% to the energy consumed in order to take into account both the vehicle manufacturing (15%) and the road construction and maintenance (35%). In the present methodology, the construction machinery footprint, when applicable, would be computed as part of the material consumption EF, in a different section.

The public infrastructure of common use, which is part of the urban, regional or national footprint calculation, it is not considered in the building construction EF.

The procedure in this case, similar to the electric power evaluation, but simplified, is as follows:

1. Calculation of fuel consumption (in litres): Generally, diesel is used.
2. Application of the conversion factor to transform this fuel consumption into units of energy (joules). Two procedures are possible: the application of the fuel energy intensity or the conversion into kg and then application of the energy intensity in MJ/kg.

    Fuel energy intensity=35 MJ/fuel liter   ...   **(12)**

3. Once the GJ are calculated, calculation of the EF in a way similar to previous sections:

$$EF_{wf} = \frac{F}{EP} \cdot e_f \qquad \qquad ... \qquad \qquad \textbf{(13)}$$

where:
EF_{wf}: EF of fuel consumption
F: Fuel consumption (GJ)
EP: Energy productivity of fuel (GJ / ha)
The oil productivity currently in Spain is 71 GJ/ha.

4. Calculation of the EF in ha/cap per year:

$$EF_N = \frac{EF}{N*t} \qquad \qquad ... \qquad \qquad \textbf{(14)}$$

where:
EF_N: EF expressed in ha/cap/year
EF: energy consumption EF expressed in ha
N: population size
T: time period

Electricity Consumption on the Construction Site

In previous methodologies [17, 18] the calculation of the EF associated with the energy consumption is performed based on the construction project budget. Polynomial formulas from official governmental publications in Spain define the energy consumption per economic sector; the energy consumption in Euros is thereby obtained.

In the present methodology, the polynomial formulae is substituted by more recent work by the authors [19], which consists of a mixed model based on theoretical and empirical data for the determination of power consumption on the construction site. The theoretical consumption is based on the indirect costs of the project. These costs refer to all those elements that cannot be attributed to a single construction unit within the building work, such as the tower cranes (which handle materials, unload trucks, *etc.* during various phases of the work). By taking the Andalusia Construction Costs Database (ACCD) and its codes as reference, all the indirect costs which consume electric power are listed in Table **14**, and the proposed coefficients necessary to determine the power demand are indicated. The engine power of common construction machinery has been analysed [20].

Table 14. Determination of the electricity consumption on the construction site.

Code	Concept	Unit	Power consumption
Machinery, equipment and tools			kWh/month
C122311	Crane	month	1,525.00
C122314	Lifting platform	month	305.00
C122315	Elevator	month	305.00
C12232	Concrete mixer	month	149.45
C12233	Machine shop of steel reinforcement	month	162.67
Ancillary and complementary facilities			Power consumption
C1231	Worksite	size	kWh/m²/year
C12311	Offices	m²	208.00
C12312	Meeting rooms	m²	208.00
C12313	Storage rooms	m²	208.00
Other electrical consumption			kWh/m²/year
C12531	Lighting of the construction site (plot)	m²	1.49
C12532	Testing facilities (a/c, heating, *etc.*)	m²	1.11

For the energy consumption of the offices and meeting rooms, an average consumption of 0.10 kW/m² for commercial and office buildings has been considered the [21]. The total annual consumption (eight hours a day, 5 days a week and 52 weeks a year) is 208.00 kWh/m^2/year.

In order to determine the lighting power consumption, the area of construction site in square metres and the minimum light necessary for a safe workplace, are assumed to be 100 lumen/m^2 [22]. Low-consumption lamps are considered with an average consumption of 70 lumen/W. Artificial light, based on the average daylight in southern Spain [23], is needed for 4 h/day on average (2h in the early morning and 2h in the evenings). Finally, lighting power consumption is calculated as 1.49 kWh/m^2.

In order to take into account the building start-up tests (A/C, heating, power supply), the electricity bills from 30 different building construction sites have been analysed. The power consumption, during the testing of the facilities at the end of the project, is 1.11 kWh per m^2 of the construction area.

Small electric equipment, such as electric hammers, screwdrivers, mixers, *etc.*, is analysed as part of the direct costs because unitary costs include small machinery in its descriptors. In case the small machines are not listed or described in the project budget, an average consumption of 19.46 kWh/m2 of construction area has been determined from empirical data of completed projects [19].

WATER

Introduction

One of the greatest causes of water pollution in the world is due to building-related activities, especially in its material manufacturing and construction work. It is necessary, therefore, to account for water consumption. However, according to the EF methodology, water consumption cannot be transformed into productive land hectares.

This section explains a method that takes into account the water consumption and its transformation into a footprint. As in previous sections, the sequential calculation of the water footprint is defined in a flowchart, (see Fig. **10**).

The EF analysis of the building construction focuses on the resources necessary to carry out the project. This analysis requires the transformation, through conversion factors, of each type of consumption into productive land hectares. Therefore, any parameter that cannot be represented in terms of productive land cannot become part of the EF indicators.

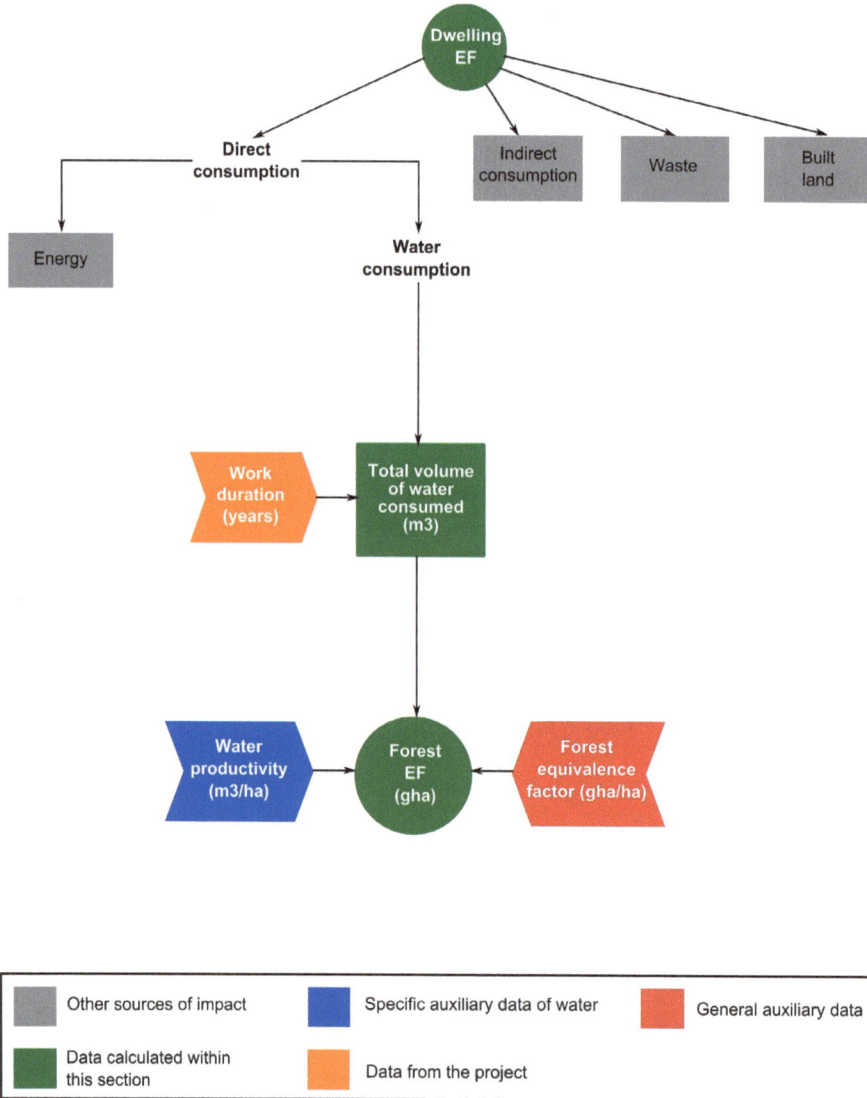

Fig. (10). Calculation of the EF, sequentially linked to the water supply.

The analysis of the water consumption EF needs to be clearly differentiated from the water footprint indicator. The water footprint of a company is defined as the total volume of freshwater that is used for the production of goods and services consumed by the company. It is expressed in units of volume, and not of productive land. It is also an indicator that comes from consumption mainly related to the residents of the building. In Fig. (**11**), the water consumption (defined as virtual water) required for the production of various food groups is represented based on data from http://virtualwater.eu/.

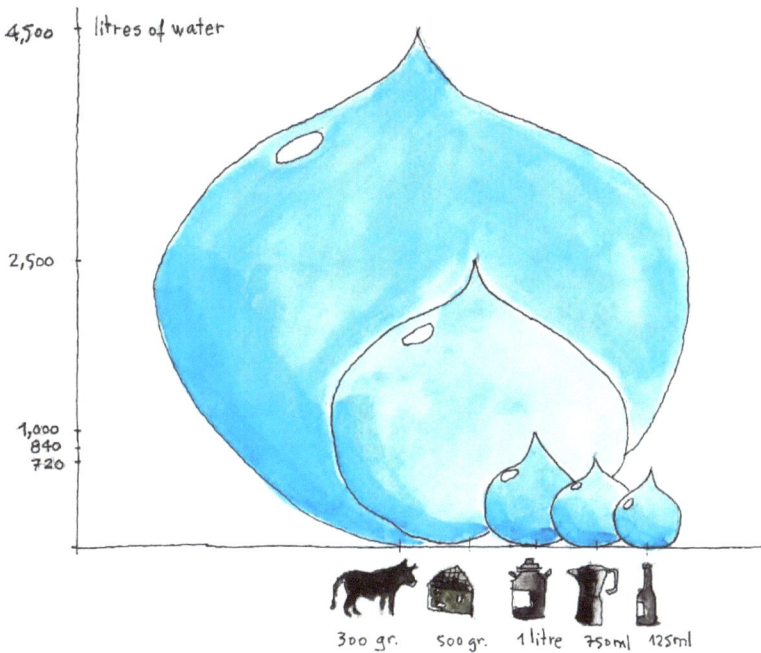

Fig. (11). The water footprint of food.

In the present analysis, it is assumed that water consumption is only due to the building construction process. As in the embodied energy concept, the water consumed contains embodied energy, that is, energy must be expended in order to attain purified water. Therefore, in order to define the embodied energy parameters associated to drinking water, it is necessary to consider the concepts included by water supply companies in the water costs per m^3. For example,

Seville's water supply company, EMASESA, includes supply, drainage, and treatment in the cost of water.

Other analyses of EF calculation (EF) related to water supply, such as that performed in the United Kingdom [15], consider the energy consumption needed to treat, supply and, where necessary, heat the water. According to this source, 1 million litres of water generate 370 kg CO_2.

Based on this data, the EF is calculated:

$$EF=370*0.00019*1.17=0.08\ ha*year \qquad ... \qquad \textbf{(15)}$$

where 0.00019 is the land needed to absorb 1 CO_2 kg, and 1.17 is the equivalence factor for forests [16].

Other authors [16], suggest that energy is consumed in pumping drinking water and treating wastewater, and that the possible leakage of supply systems should also be considered. That is, energy is required for the transfer, treatment, and supply of water, and for the treatment in the sewage plant. The EF calculation by components for the city of London [24] estimates 0.002 gha/cap. This amount is based on the consumption per dwelling. Although the calculation procedure is not defined, this result serves as a reference for future comparison.

When water consumption is linked to energy consumption, its footprint is assigned to forest land, which absorbs CO_2 emissions from fossil fuels that drive supply, drainage and wastewater treatment systems.

Liverpool also analyses the footprint related to the water supply [25]. The energy required for $41.6*10^6$ litres is 10.4 GWh, which implies a footprint of 0.02 ha/cap. Wastewater treatment is 268.7 kWh per 10^6 litres.

Wackernagel and Nerea calculate the EFs of various sectors, including that of water consumption [11]. This analysis considers the forest as a water producer, and hence the consumption of this resource is included in the forest land consumption. The water productivity (m^3/ha/year) studies are based on wetland forest which produces 1,500 m^3 of fresh water per hectare per year. The forest water production can be secondary, which often leads to the omission of this footprint calculation, although, in many areas, it is already considered a forest primary use and its corresponding environmental footprint is therefore calculated.

In the various calculation approaches described above, the calculations are associated with the building usage but not with their actual construction, (see examples of London and Liverpool).

Two different approaches are defined: the first considers energy consumption (embodied energy) linked to water supply (example of UK, London or Liverpool), and the second, uses Wackernagel and Nerea´s method, where the forest is considered as a water producer. In both cases, the footprint is calculated in terms of forest land.

Water Consumption

In the present calculation procedure, the forest is considered as a water producer, since the resulting EF calculation is more direct. On the one hand, water consumption is not transformed into other units, and on the other hand, it is more interesting to link water consumption to its production processes, rather than to a CO_2-emission process.

The water consumption for the building construction under study can be estimated with representative coefficients. The methodology used to determine the footprint linked to the water supply for the case of a residential building construction is summarized in Fig. (**10**). The steps are:

1. Determine water consumption in works of similar dimensions to the one to be analysed.
2. Define the average water consumption of the construction work being analysed, with the data obtained in point 1, based on the project size (see Table **15**).
3. Determine the EF, according to Wackernagel and Nerea´s coefficients [11]:

$$EF = \frac{W}{FP} \qquad \cdots \qquad (16)$$

where:
EF: EF of water consumption
W: Water consumption (GJ)
FP: Forest productivity (GJ/ha)

and where the forest productivity, as mentioned above, is taken as 1,500 $m^3/ha/year$; this value must be updated periodically.

4. Apply the value obtained from the EF equivalence factor corresponding to the pro-ductive land considered, in this case the forest land Table **3**, as stated in the following equation:

$$EF_{ww} = \frac{W}{FP} \cdot e_f \qquad \cdots \qquad (17)$$

where:

EF_{ww}: weighted water footprint (gha)

e_f: equivalence factor for forests, (currently calculated at 1.34)

The water consumption data from Seville's water supply company from ninety projects has been analysed, and the corresponding average water consumption per project type has been determined. The results are summarized in Table **15** [18].

Table 15. Water consumption per dwelling type [18].

No. of floors above ground level	Sample size	Average consumption ($m^3/year/m^2$)
1	22	0.29
2	31	0.24
3	21	0.13
4	15	0.15
5	3	0.15
10	2	0.07

It is important to clarify that the water consumption on the construction site cannot be determined through direct or indirect costs analysis. The equation presented by Domenech, previously defined by Wackernagel and Nerea, provides a good approximation for the calculation of this footprint.

Data on conversion factors should be taken into account in EF calculations. In this case, the equivalence factor for forests and the forest productivity for the generation of fresh water are both needed. If this data is corrected or modified, then the calculation of the footprint is also affected.

Other Considerations

In London's footprint, the water consumption per dwelling is 150 litres per family per day, based on an average family of 2.7 members that consumes 47 m^3 of water per year per member.

Other bibliographic sources consider a standard consumption per habitant (litres per person per day) of 150 litres [26]. According to the latest data available, the average consumption of Barcelona's metropolitan area is 128 litres per person per day [27].

CONCLUDING REMARKS

The impact of energy and water, a direct consumption on the construction site, is included in the EF calculation. The energy is obtained from fuel and electricity. The transformation of these two types of consumption into EF values is performed through the existing EF methodology, although certain procedures have to be adapted to the building sector.

Combustion engines of heavy machinery and power plants consume petrol or diesel. In project budgets, the machine description and the working hours are described separately, which allows the calculation of its fuel demand as a direct calculation.

The electric power consumed on the construction site is calculated from the indirect costs of the budget. These costs refer to all those elements that cannot be attributed to a single construction unit within the building works, such as the tower cranes (which move materials, unload trucks, *etc.* during various phases of the work, such as those based on the foundation, structure, finishes, *etc.*). All the indirect costs which consume electric power are listed and transformed into power consumption. Various coefficients have been defined in order to perform the calculation.

The conversion of energy consumption into productive territory considers forest land as the productive land necessary for the absorption of CO_2 emissions resulting from the burning of fuel. In the energy footprint of the building, an average absorption factor is applied. The energy productivity is obtained from absorption and emission factors.

Secondly, the water supply EF is evaluated. In order to determine the direct water consumption on the construction site, empirical data from 90 construction projects has been analysed, and coefficients depending on the building type have been determined.

Generally, all EF studies obviate this aspect due to the intrinsic difficulty of transforming water consumption data into a quantity of consumed land. The forest productivity is employed, which is taken as 1,500 m³/ha/year.

CONFLICT OF INTEREST

The author confirms that this chapter has no conflict of interest.

ACKNOWLEDGEMENT

Ministry of Innovation and Science, through the concession of the R+D+I project: Evaluation of the EF of construction in the residential sector in Spain. (EVAHLED). 2012-2014. *Ministerio de Innovación y Ciencia, por la concesión del Proyecto I+D+i: Evaluación de la huella ecológica de la edificación en el sector residencial en España (EVAHLED). 2012-2014.*

REFERENCES

[1]   G. Acosta Bono, J. González Daimiel, M. Calvo Salazar, and F. Sancho Royo, *Estimación de la Huella Ecológica en Andalucía y Aplicación a la Aglomeración Urbana de Sevilla (Estimation of the Ecological Footprint in Andalusia and Application to the Urban Agglomeration of Seville)*. Dirección General de Ordenación del Territorio y Urbanismo, Consejería de Obras Públicas de la Junta de Andalucía: Seville, Spain, 2001. [Online] Available: http://hdl.handle.net/10326/974 [Accessed Sept 18, 2014].

[2]   IDAE, *ES6: Energy Efficiency and Renewable Energy, num1*. IDAE: Bulletin Madrid, Spain, 2000.

[3]   IDAE, *Renewable Energy Plan in Spain 2005-2010*. IDAE: Madrid, Spain, 2005.

[4]   Spain Ministry of Industry, *Energy in Spain 2006*. Madrid, Spain, 2007.

[5]   Spain MI (Ministry of Industry), *Estructura de generaciOn eléctrica en España. La Energía en España 2007 (Structure of generation of electricity in Spain. Energy in Spain 2007)*. 2008. [Online] Available: http://www.aven.es/pdf/la_energia_en_espana_2007.pdf [Accessed Sept 18, 2014].

[6]   A. Cuchí, and I. López Caballero, *Informe MIES (MIES Report: An Approach to Environmental Impact of the School of Architecture of the Vallès)*. Universidad Politécnica de Cataluña (UPC):

Barcelona, Spain, 1999.

[7]   F. Mañá, and A. Cuchí, *Parámetros de sostenibilidad (Sustainability parameters)*. ITEC: Barcelona, Spain, 2003.

[8]   IDAE, *ES5: Plan for Renewable Energy Development. Appendix I: Units and Conversion Factors.* IDAE: Madrid, Spain, 2000.

[9]   Instituto para la Diversificación y Ahorro de la Energía (IDAE), *Guía Práctica de la Energía: Consumo Eficiente y Responsable (Practical Energy Guide: Efficient and Responsible Consumption)*IDEA: Madrid ,Spain, . [Online] Available: http://www.idae.es/index.php/mod. documentos/mem.descarga?file=/documentos_11406_Guia_Practica_Energia_3ed_A2010_509f8287. pdf. [Accessed Sept 18, 2014].

[10]   N. Casado, J.M. González, J.I. Llorens, F. Maña, P. Martorell, and A. Puig-Pey, *La enseñanza de la arquitectura y del medio ambiente (The Teaching of Architecture and Environment)*. ITEC: Barcelona, Spain, 1997.

[11]   J.L. Domenech Quesada, *Huella ecológica y desarrollo sostenible (Ecological Footprint and Sustainable Development)*. AENOR: Madrid, Spain, 2007.

[12]   M. Wackernagel, and W. Rees, *Our Ecological Footprint: Reducing Human Impact on the Earth*. New Society: British Columbia, Gabriola Island., 1996.

[13]   M.E. Figueroa Clemente, and S. Redondo Gómez, *Los Sumideros Naturales de CO$_2$: una Estrategia Sostenible entre el Cambio Climático y el Protocolo de Kyoto. (Natural sinks of CO$_2$: A Sustainable Strategy between Climate Change and the Kyoto Protocol)*. Universidad de Sevilla: Seville, Spain, 2007.

[14]   P. Coto Millán, J.L. Domenech Quesada, and I. Mateo Mantecón, "Corporate Ecological Footprint: New Conversion Factors", *Res. Lett. Ecol.,* 2008.
[http://dx.doi.org/10.1155/2008/415934]

[15]   N. Chambers, C. Simmons, and M. Wackernagel, *Sharing Nature's Interest: Ecological Footprints as an Indicator of Sustainability*. Sterling Earthscan: London, Great Britain, 2004.

[16]   T. Wiedmann, J. Barrett, and N.Cherrett N, "Sustainability Rating for Homes", In: *The Ecological Footprint Component*. Stockholm Environment Institute: York, Great Britain, 2003. [Online] Available: http://www.sei.se/index.php?section=implement&page=publications. [Accessed Sept 10, 2014].

[17]   J. Solís-Guzmán, M. Marrero, and A. Ramírez-de-Arellano, "Methodology for determining the ecological footprint of the construction of residential buildings in Andalusia (Spain)", *Ecol. Indic.,* vol. 25, pp. 239-249, 2013.
[http://dx.doi.org/10.1016/j.ecolind.2012.10.008]

[18]   P. González-Vallejo, M. Marrero, and J. Solís-Guzmán, "The ecological footprint of dwelling construction in Spain", *Ecol. Indic.,* vol. 52, pp. 75-84, 2015.
[http://dx.doi.org/10.1016/j.ecolind.2014.11.016]

[19]   M. Marrero, A. Freire-Guerrero, J. Solís-Guzmán, and C. Rivero-Camacho, "Estudio de la huella ecológica de la transformación del uso del suelo", *Seguridad y Medio Ambiente,* vol. 136, pp. 6-14, 2014. ISSN: 1888-5438. http://www.mapfre.com/documentacion/publico/i18n/catalogo_imagenes/

grupo.cmd?path=1081726. [Accessed Feb 13, 2015].

[20]   A. Freire-Guerrero, and M. Marrero, "Analysis of the ecological footprint produced by machinery in construction", In: *World Sustainable Buildings Congress*, Barcelona, Spain, 2014.

[21]   Spain MS (Ministry of Science), *Real Decreto 842/ 2002, por el que se aprueba el Reglamento electrotécnico para baja tensión (Royal Decree 842/2002, low voltage electrical regulation)*, 2002.

[22]   Spain ML (Ministry of Labor), *Real Decreto 486/1997, de 14 de abril, por el que se establecen las disposiciones mínimas de seguridad y salud en los lugares de trabajo. (Royal Decree 486/1997, of minimum safety and health in the workplace)*, 1997.

[23]   Spain MH (Ministry of Housing), *Código Técnico de la Edificación (Building Technical Code)*, Madrid, Spain, 2006. http://www.codigotecnico.org/web/ [Accessed 1st December 2013].

[24]   M. Nye, and Y. Rydin, "The contribution of ecological footprinting to planning policy development: using REAP to evaluate housing policies for sustainable construction", *Environ. Plann. B Plann. Des.*, vol. 35, no. 2, pp. 227-247, 2008.
[http://dx.doi.org/10.1068/b3379]

[25]   J. Barrett, and A. Scott, *An Ecological Footprint of Liverpool: Developing Sustainable Scenarios. A Detailed Examination of Ecological Sustainability*. Stockholm Environment Institute: York, Great Britain, 2001.

[26]   A. Cuchí, *Arquitectura y sostenibilidad (Architecture and Sustainability)*. Universidad Politécnica de Cataluña (UPC): Barcelona, Spain, 2005.

[27]   T. Solanas, and J. Herreros, *Vivienda y sostenibilidad en España. Volume 2: Colectiva (Housing and Sustainability in Spain. Volume 2: Collective)*. Gustavo Gili: Barcelona, Spain, 2008.

CHAPTER 4

Indirect Consumption: Manpower and Construction Materials

Abstract: This chapter analyses the impact sources that consume resources indirectly, that is, the impact is caused not by the source, but by its components. For this study, we focus on two of these components: manpower and material consumption, both of great importance in the EF calculation.

First, the manpower consumption is studied by focusing on the most determinant aspects of its impact: food and mobility. The transformation of these consumptions into EF values is performed by previously documented processes which are adapted to the specific characteristics of the building sector. Second, the EF associated to the consumption of construction materials is evaluated during the building execution process; which takes into account the energy consumption deriving from the manufacture, transport and installation of each of the materials used in the construction of buildings.

Keywords: Absorption factor, Construction materials, Conversion factors, Dwelling construction, Ecological footprint, Electricity, Embodied energy, Emission factor, Forest productivity, Fuel productivity factor, Manpower, Mobility, Natural productivity, Productive land, Water consumption.

INTRODUCTION

This chapter analyses the impact sources that consume resources indirectly, that is, the impact is caused not by the source, but by its components. For this study, we focus on two of these components: manpower and construction material consumption, both of great importance in the EF calculation.

First, the manpower consumption is studied by focusing on the most determinant aspects of its impact: food and mobility. The transformation of these consumptions into EF values is performed by previously documented processes which are adapted to the specific characteristics of the building sector. Second, the EF associated to the consumption of building materials is evaluated during the building execution process. This section analyses the energy consumption deriving from the manufacture, transport and installation of each of the materials used in the construction of buildings.

MANPOWER

The consumption associated to manpower is divided into that arising from the workers' transportation and that caused by the food consumption on the construction site.

Food

The methodology follows the sequence described in Fig. (**12**). To calculate the EF of the building sector, it is necessary to take into account the consumption caused by all actors involved in the execution of the work. Therefore, food consumption is also an impact source to consider, although its impact is indirect. Food consumption, in all EF analyses at consumer level, appears as a parameter of study, and is considered in the consumption rates of the building sector.

Table 16. The EF of food.

Food	EF (gha per year per tonne)
Cereal	1.7 - 2.8
Pulses	3.6 - 4.4
Vegetables	0.3 - 0.6
Meat	6.9 (pasture) - 14.6 (animals in captivity)
Milk	1.1 - 1.9
Fish	4.5 - 6.6
Fruit	0.5 - 0.6

Consumer goods in general and food in particular, are usually expressed in tonnes consumed, and then transformed into hectares to obtain EF rates. In fact, in the analysis of the EF by component for regions or organizations, one of the typical components of study is food consumption, expressed in tonnes [1].

A summary of the results of this analysis is shown in Table **16**, and graphically in Fig. (**13**). It shows a list of the most important foods, classified into large groups,

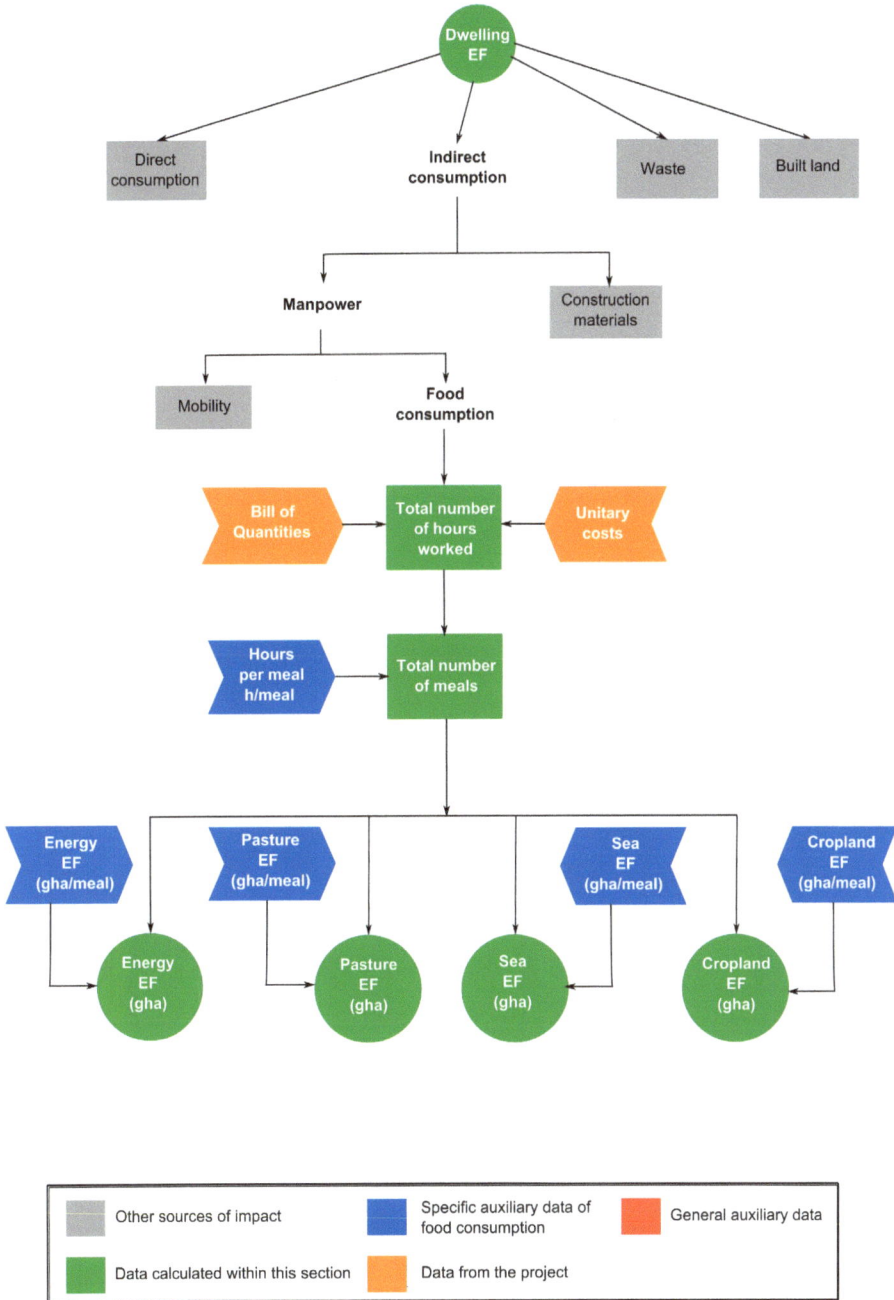

Fig. (12). Sequential calculation of footprint associated with food consumption.

with the corresponding impact on the territories generating each of them.

These impacts are given in gha per year per tonne. The gha unit is used because it refers to the global productivity of each of the territories or lands that generates consumer goods. For example, to ascertain the EF of cereal crops, global average productivity of cereal crops is used. The calculations include the percentages of transport, processing and energy required.

In food production, the productivity is specifically called *natural productivity*, and is defined as the amount of territory (ha) required to produce 1 tonne of resources (food).

Fig. (13). The EF of different kinds of food [2].

If a community annually eats 5,000 kg of tomatoes, and the average natural productivity of tomatoes is 5,000 kg/ha, then that community consumes the equivalent of 1 hectare of cropland, which is its corresponding footprint for that land type.

In order to calculate the food EF, the following expression is used:

$$EF = \frac{C}{NP} \qquad \ldots \qquad (18)$$

where:
EF: food EF (ha)
C: consumption (t)
NP: natural productivity (t/ha)

In addition to quantifying the productivity of each type of land, it is necessary to assign a type of productive land to each of the consumer goods. Some examples are:

- The potato consumption footprint is assigned to cropland.
- The fish consumption footprint is assigned to productive sea.
- The wood consumption footprint is assigned to forest land (forests).
- The meat consumption footprint is assigned to pasture, since this is typically the kind of land in which the different types of meat-producing animals are located.

The footprint of the production of raw material is obtain from its natural productivity, but foods processed from raw materials include an energy footprint, that is, the energy spent on processing foods from raw materials, in the form of the consumption of fossil fuel:

$$EF_f = \frac{C \cdot EI}{EP} \qquad \ldots \qquad (19)$$

where:
EF_f: energy footprint (fossil fuel) of food (ha)
C: consumption (t)
EI: energy intensity (GJ/t)
EP: oil productivity (GJ/ha)

The footprint calculations on different types of food focus on consumption data available (in tonnes) and productivity. The footprint provided by Wackernagel and Nerea associated with agricultural resources is provided in Table **17**, which includes all process inputs, such as chemical fertilizers, pesticides, and treatments [3]. That is, the calculation of the energy consumption of each food type includes all previous treatment previous to harvesting.

Table 17. Energy intensity and natural productivity of food.

Food	Energy intensity (GJ/t)	Natural productivity (t/ha/year)
Meats	80	0.033
Seafood	100	0.029
Cereals, flour, pasta, rice, bread	15	2.264
Drinks (juice, wine)	7	22.500
Vegetables	10	6.730
Sweet food	15	4.893
Oils and greases	40	1.485
Dairy	37	0.276
Coffees / Teas	75	0.566

The foods which produce high footprints are those of fish and meat, due to their low productivity. Two columns are defined: one refers to the natural productivity values, as explained above, and the other column to the energy intensity involved in the processing.

Energy intensity measures the energy consumption (GJ) for a tonne of each of the agricultural resources to be rendered readily available to consumers, that is to say, ready for consumption. This parameter enables the consumption of various resources to be linked with the fossil footprint indicator. Energy intensity, natural productivity, and annual consumption of each of the resources, determine the EF generated by the food intake of the population under study.

In the methodology for calculating the footprint associated with food intake, a group of foods are defined to form a meal. Domenech [3] defines the average menu of the works cafeteria of a company, summarized in Table **18**. Once the percentages, from a budget point of view, are defined, the last column converts the values into tonnes, the values are defined by the Spanish Foreign Trade statistics [3].

A new methodology is proposed for the calculation of the EF generated by a worker's meal, where data from the Food and Agriculture Organization of the United Nations (FAO) [4] and the Global Footprint Network [5] are analysed and combined (using weighted averages) in order to determine a daily meal footprint. In the previous methodology, Solis-Guzman *et al.* [6] and Gonzalez-Vallejo *et al.* [7] based on Domenech´s work [3], calculate the EF of a worker's meal from economic data and evaluated a 10-Euro cafeteria menu. In the new approach, the menu is defined based on FAO data for Spain that can be grouped into a general classification, (see Table **19**). Their corresponding intensity factors are published by the Global Footprint Network [8]. The energy intensity is obtained from Wackernagel and Rees [8], as in the original methodology. Finally, the amounts of food consumed daily are not defined by cost but by weight, (Table **19**), which generates an EF that is 10 times lower than in previous work, (Table **20**).

Table 18. Composition of a company menu and conversion into tonnes.

Food	%		%	Conversion (t/1,000 €)
Meats	25	Poultry	25	0.65
		Pork	25	
		Grain cattle	25	
		Grazing cattle	25	
Fish	25			0.50
Cereals	12			4.69
Drinks	10			0.34
Beans and potatoes	8			1.45
Sweet food	6			0.70
Oils	5			0.71
Dairy	5			0.93
Coffees / Teas	4			0.54

In addition to the information shown, it must be borne in mind that arable land is inexistent in built land, and therefore food consumption is always an external source, and the footprint is applied to the agricultural and fishery resources section. If later, during the useful life of the building, some crop type is grown in the area that was previously used as built land, then that land should be counted as natural capital, reducing the footprint of this class.

Certain authors [3] affirm that manpower should have no impact on the EF of the company to be analysed, because is assumed that each and every personal consumption (maintenance, travel to work, *etc.*) belong to the individual footprint, regional or national, but not to the corporate footprint.

Table 19. The daily food consumption in Spain and its corresponding production and energy intensities.

Item	Consumed (kg/day)[4]	(%)	Intensity (gha/t) [5]			Energy (GJ/t) [8]
			crops	pastures	sea	
Cereals - Excluding barley for beer	0.30	7.72	10.47			10
Starchy Roots	0.18	4.73	0.17			5
Sugar Crops	0.25	6.41	0.06			15
Sugar & Sweeteners	0.07	1.75	0.37			15
Pulses	0.03	0.67	1.53			10
Tree nuts	0.02	0.47	2.02			10
Oil crops	0.71	18.45	1.63			10
Vegetable Oils	0.09	2.37	7.32			40
Vegetables	0.41	10.57	0.17			100
Fruits - Excluding those for wine	0.55	14.36	0.30			10
Stimulants	0.02	0.54	4.72			75
Spices	0.00	0.03	2.75			75
Alcoholic Beverages	0.30	7.87	0.29			15
Eggs	0.04	1.01	1.65			65
Meat	0.26	6.63	1.77	2.28		80
Animal fats	0.01	0.35	2.78	4.69		40
Milk - Excluding that for Butter	0.49	12.81	0.57	1.03		10
Fish, Seafood	0.12	3.02	3.22		8.14	100

However, the construction process considers manpower as a resource to be used in the execution of the building, and hence it is essential to include it as generating an impact factor. For our case study, that is to say, in the construction phase of residential buildings, only the maintenance footprint is considered, in other words, the footprint associated with food consumption. This is estimated in terms of:

- Time (work duration in days), the working time is taken as 8 hours per day, 40 hours per week.
- Number of employees
- Number of meals on the construction site
- Food types

The project budget is used to calculate the food footprint, since construction projects include the manpower cost per hour worked. From the cost of manpower per hour (€/h), the total cost of manpower in hours is obtained for the construction project under study. Finally, the food consumption associated to those hours is analysed, using the vales given in Table **18**.

Table 20. The EF of the daily food consumption in Spain per year and person.

Crops EF (10-3 gha)	Pastures EF (10-3 gha)	Sea EF (10-3 gha)	Energy EF (10-3 gha)	Source
8.50	23.40	15.90	7.20	[7]
3.89	1.15	0.94	0.17	New methodology

From the project budget, the actual manpower used in the construction is obtained, for which the steps include:

1. Use the project budget in order to determine the impact of manpower in monetary terms.
2. Convert € into hours, by using the rate of manpower cost in €/h.
3. For the conversion of these hours into ecological cost, base the ratio ha/h of food consumption on the productivity data of Table **17**, and on the equations for pro-ductivity, energy intensity, and consumption.
4. Assign each of the footprints to their corresponding territory. Finally, several types of footprints for food consumption are determined: cropland, pastures, productive sea, and forest.

Indirect costs associated with manpower are also included, such as those corresponding to supervisors, foremen, technicians, administration, auxiliary staff, (which are not part of the direct costs of the project).

Mobility

The methodology to calculate the mobility EF follows the sequence described in Fig. (**14**). According to IDAE data [9], cars consume 12% of the total energy in Spain and are responsible for 40% of the energy consumption in road transportation; the average annual family expenditure in car fuel is 1,200 €.

Table 21. Transport consumptions in Spain, energy units per passenger per kilometre.

Means of transport	Energy per km per passenger
Bus	1
Car	2.9
Plane	12.1
High-speed train	1.8
Railway	2.6

According to Table **21**, except for air travel, the car represents the traveller's most inefficient transport, and generates a huge ecological impact [3]. During intercity trips, a car consumes per passenger and kilometre almost 3 times that of a bus, and this difference is even higher in urban areas. Other EF calculations, associated with mobility, are shown in Tables **22** and **23** [9, 10]. Another source calculates the emissions generated by vehicle transport as 133 g CO_2/km travelled [11].

A relevant study has been carried out in Catalonia, Spain [12], where student mobility between the campus and the city centre is analysed. The study considers transport type, number of miles covered annually, energy consumption, and CO_2 emissions. This data is needed in order to calculate the mobility impact in terms of productive land hectares. The results concerning the energy consumed in transporting students to and from the university are given in Table **24**.

Table 22. Coefficients of fuel consumption of cars in Spain.

Fuel	Consumption (liters/100 km)	CO_2 Emissions (kg/liter)
Gasoline	7.40	2.35
Gasoil/diesel	6.04	2.60

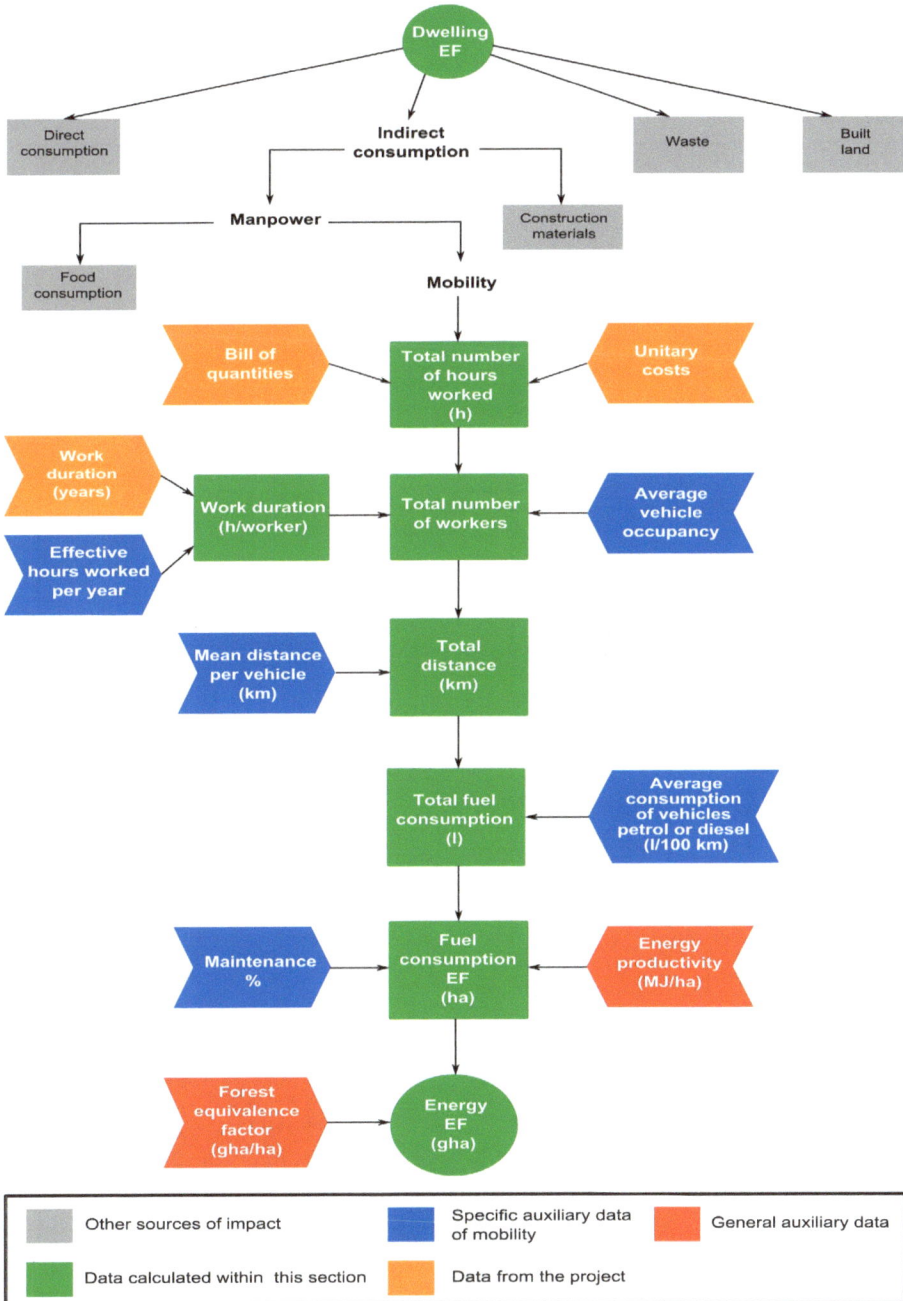

Fig. (14). Calculation of the mobility footprint.

Table 23. Distance travelled per passenger with 1 liter of fuel.

Means of transport	km/liter per passenger
Bus	39.50
Car	18.60
Railway	48

Table 24. Results of the university mobility study.

Transport	Persons (%)	Annual mileage	Energy consumed (MJ)	Impact (%)	CO_2 Emissions (kg)
Car	44.9	3,124,980	8,565,800	89.8	642,432
Train	37.4	2,610,894	858,036	9.6	68,536
Bus	2.8	140,533	54,808	0.6	4,110.6

Other relevant mobility studies originate from the UK, perhaps since Wackernagel and Nerea's methodology was promptly applied in various populations, leading to infor-mation being made available in this field. In addition, we also observe that the British public entities are currently using EF indicator as a reference for a wide variety of reports, with useful calculations and procedures.

Various studies performed in Britain evaluate the EF of transportation by car; all of these studies follow the methodology defined by Wackernagel and Nerea, and the consequent data has been found useful in the proposed methodology. The first study was conducted to determine the EF of trips by car in the UK, whose most relevant data is summarized in Table **25** [1].

Table 25. EF of car journeys in the UK, the demand of 10,000 passengers per kilometre per year.

Component	Demand	Land type	Equivalence factor
Oil consumption	0.094 l/km (0.22 kg CO_2/km)	CO_2 absorption	1.17
Roads	2,581,747 ha	Constructed area	2.83
Cars with respect to all vehicles on road	86%		
Annual trips by car	362,400,000,000 km		
Occupation per car	1.6 people		

According to this study, in order to calculate the EF of the trips by car, we must consider:

1. Fuel consumption: this quantity can be expressed in many different ways: l/km, kg CO_2/km, *etc*. It is measured in terms of the territory necessary for the CO_2 absorption, or in other words, as forest land.
2. The energy consumption during the vehicle manufacture and maintenance is consi-dered to be equivalent to 15% extra in the fuel consumption [1].
3. Energy consumption for road construction and maintenance is considered to be equi-valent to 30-35% extra fuel consumption [1].
4. The land used to build roads is considered as a footprint of directly occupied land. It is necessary to ascertain the surface area occupied by the roads, the number of cars on the roads, and the annual car trips which take place in the territory (expressed in km).

This data obtains the footprint in terms of CO_2 absorption (points 1, 2, and 3) and in terms of built area (point 4). The equivalence factors transform the local productive hectares into global hectares.

The EF calculation is not expressed in annual hectares per car per kilometre but is determined per passenger per kilometre, by employing the car occupancy data.

The calculations of Table **25** are expressed in (m^2) instead of (ha), and hence in order to determine point 4 (built land) the calculation involves the following:

$$\frac{[2,581,747(1)*0.86(2)*10000(4)]}{[362,400,000,000(3)]} = 0.06 \cdots \quad (20)$$

$$0.06 * 2.8 = 0.17 \, m^2 \text{ per car per km per year} \quad \cdots \quad (21)$$

where:
(1): total UK road area is given in hectares
(2): car use percentage with respect to all vehicles on road
(3): distance covered by car trips made in the UK per year (expressed in km)
(4): conversion factor of hectares into km^2

The equivalence factor is 2.8.

Other studies into the calculation of the EF of car trips in the UK consider similar parameters and methodology [13]. Cars represent 86% of all road users, and have an average occupancy of 1.6 persons per car. The total EF is established as the sum of the oil consumption footprint plus the road usage footprint.

Similar results are determined in another study, also in the United Kingdom [14], based on Chambers and Simmons's EF analysis by components. Through the analysis of the component structure, the so-called Eco-Index Methodology is developed, used primarily to determine the domestic consumption of housing by means of a computer application, ECOCAL [15]. Among other results, this application enables the calculation of the fuel consumption associated to transportation by car in the United Kingdom. The conversion factors came from Chambers *et al.* [1]. For the case of car trips, the application uses similar figures see Table **26** to those mentioned in the previous paragraph.

Table 26. Values of mobility EF.

Transport	EF (10^{-5} ha/passenger/km)
Car	3.64
Motorbike	2.93
Train	2.41
Plane	7.35
Bus	2.93

The mobility footprint is composed of several terms and, in the present analysis, a specific procedure is proposed. Undoubtedly, in order to ascertain the impact generated by the workers' transportation to the construction site, the fuel consumption of cars needs to be determined (this being the most likely means of transport used).

Regarding the fuel consumption during the vehicle manufacturing and maintenance, this is considered as 10% of fuel over cost, since each car has other uses which cannot be assigned directly to the building work. Logically the energy consumption during the manufacturing of the car is not included in the material consumption section.

Table 27. Coefficients for the calculation of CO_2 emissions (CRAG).

Consumption	Conversion factor (kgCO$_2$/unit)
(l) Gasoline liters	2.3
(l) Diesel liters	2.7
(km) travelled in small gasoline-fuelled car (<1.4 l)	0.17
(km) travelled in medium-sized gasoline-fuelled car (1.4-2.1 l)	0.22
(km) travelled in large gasoline-fuelled car (> 2.1 l)	0.30
(km) small diesel-fuelled car (< 1.7 l)	0.15
(km) medium-sized diesel-fuelled car (1.7-2 l)	0.19
(km) large diesel-fuelled car (> 2 l)	0.26

The footprint of public infrastructures, both of the fuel consumption during the cons-truction and maintenance of the roads, and of the land surface directly occupied by the road infrastructure, is included in the regional or national urban footprint calculation, but not in the building project, where the consumption derived exclusively from the building construction process is considered, following the analysis of other authors [3]. Therefore, the hypothesis excludes the use of public infrastructures from the present footprint calculation.

The fuel consumption data is necessary to determine the mobility footprint. This data can be obtained from studies by British public institutions (DEFRA and CRAG), which calculate CO_2 emissions in the UK. In the case of CRAG [16], the emissions are given depending on fuel consumption ranges, (see Table **27**).

where CO_2 emissions are calculated according to the following expression:

$$E=C * Fc \qquad \text{...} \qquad (22)$$

where:
E: CO_2 emissions (kg)
C: consumption (consumption unit, Table **27**)
Fc: conversion factor

DEFRA [17] is a similar calculation tool of CO_2 emissions. In Table **28**, its corresponding calculations for car journeys appear. In the present methodology, following the flowchart in Fig. (**14**), the results in Table **24** are combined with recent analysis [18].

Table 28. Coefficients for the calculation of CO_2 emissions (DEFRA).

Transport/consumption	kg CO_2 / consumption unit
(l) Gasoline	2.32
(l) Diesel	2.63
(kWh) Electric vehicle	0.52
(km) Hybrid car (medium-sized)	0.13
(km) Hybrid car (large)	0.22

In the present methodology the mobility is calculated following (Fig. **14**). The final assumptions are:

1. Private vehicles are the means of transportation of the operators, and the work site is situated in a remote area, far from the urban area, which renders travel by public transport infeasible.
2. The average distance travelled by the operator is 15 km each way.
3. The average occupancy per vehicle is 4 people per vehicle.
4. Calculation of the fuel consumption is based on (Table **23**). An extra consumption of 10% is added in order to include the vehicle maintenance.
5. Calculation of CO_2 emissions is obtained from Tables **22 - 28**, and the fuel consumption is calculated following the same procedure described in the energy section.
6. The mobility footprint is determined following the procedure described in the energy section.

Finally, as in the case of food consumption, the mobility EF is considered an indirect consumption.

CONSTRUCTION MATERIALS

As in previous sections, the methodology is described in a flowchart, (see Fig. **15**). The consumption of building materials involves several characteristics which lead to complications in the calculation of its footprint, mainly because the documented procedures for calculating the energy consumption resulting from the manufacture, transportation and installation of these materials are generally inaccessible or insufficiently justified.

Before establishing the calculation methodology of the footprint of construction materials, the procedure for the evaluation of energy consumption must be clarified, thereby making it necessary to explain the concept of embodied energy.

Embodied Energy

In order to adequately explain the concept of embodied energy, it is necessary to previously define the phases of the life cycle of a material used in construction. The concept "from cradle to gate", refers to the material manufacturing process, from its extraction as raw material, through manufacturing processes and transportation in the processing plants, up to the point where the manufactured product is ready for delivery at the gates of the factory. "From cradle to site" also includes the transport to the construction site. "From cradle to grave", includes all phases of the life cycle of the material: manufacture, use, disposal and material recycling, (see Fig. **16**). Finally, the concept "from cradle to cradle" has to do with the materials that lengthen the life cycle of the dwelling, by being again part of their manufacturing process, (see Fig. **17**).

There are many ways to approach the life cycle of a product, therefore rendering it necessary to define it properly in the present methodology. In the construction process, the products studied are from *cradle to gate* or *cradle to site*; other life cycle phases, such as use, demolition and recycling of these materials, are not considered.

There are also several ways of defining the embodied energy concept. According to Bjorn Berge [19], the embodied energy of a product includes all the energy consumed by processes from the cradle to the gate. This energy includes the

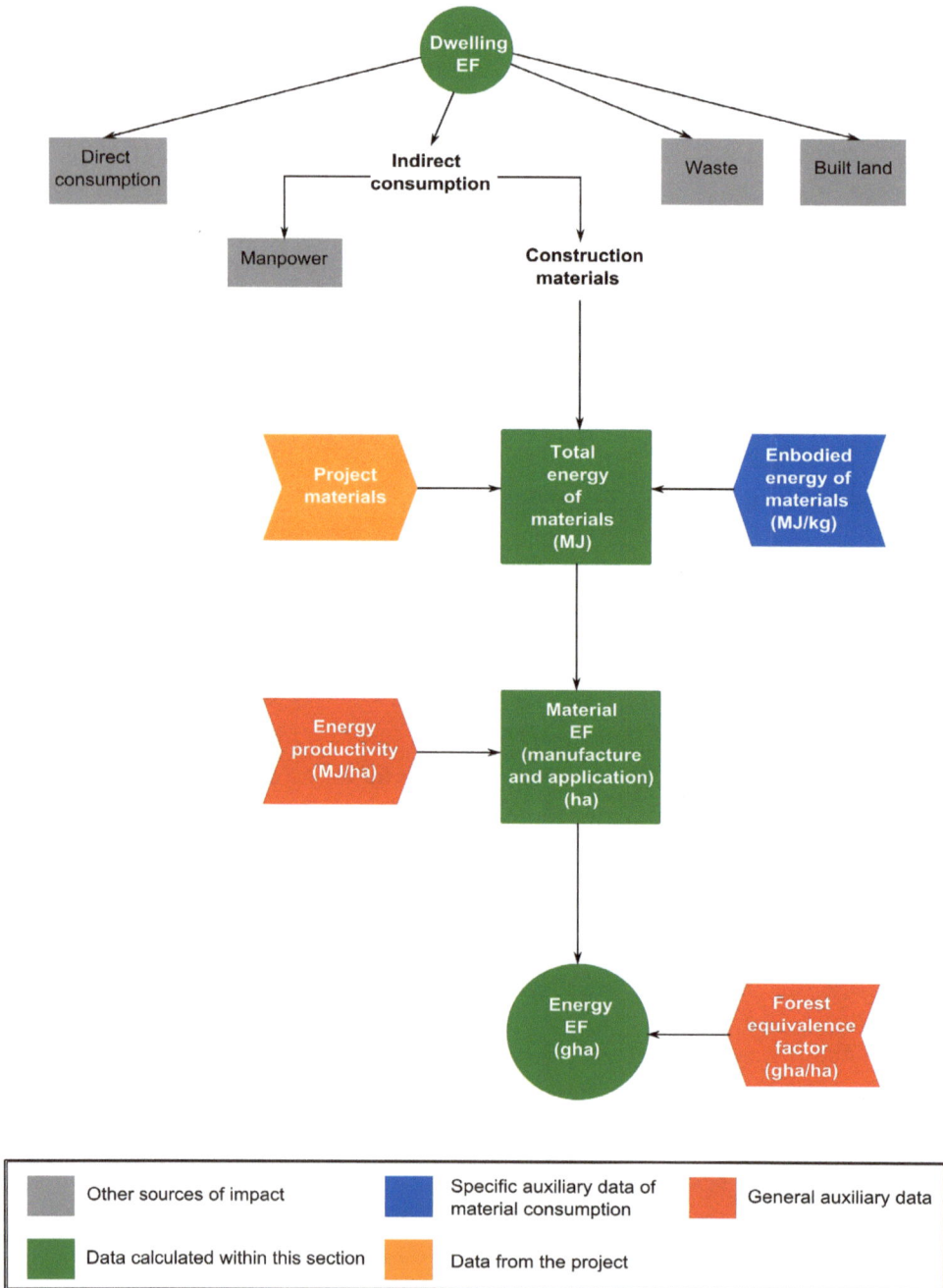

Fig. (15). Methodology for calculating the footprint of construction materials.

energy released in the product combustion process, in the case where it is incinerated at the end of its life. Accordingly, this embodied energy represents 85-95% of the total energy spent on that product which is part of the building. The remaining 5-15% refers to the transport, installation, maintenance and demolition that take place in the building construction process. Similar reasoning can be found in other studies [20].

Fig. (16). The life cycle of construction materials.

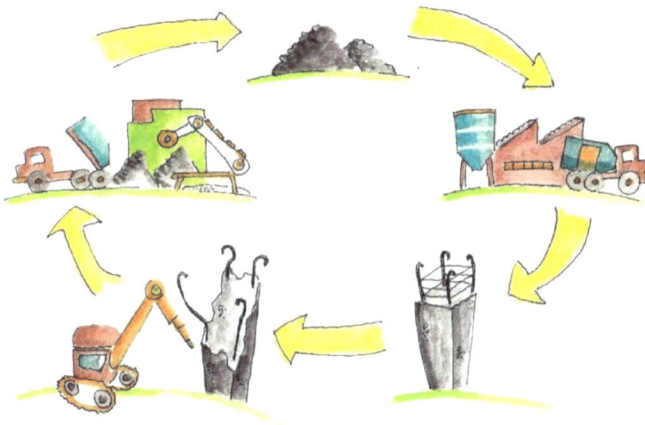

Fig. (17). Concrete production from "cradle to cradle".

In yet another approach, the embodied energy is defined as one that includes production, transport and installation of materials which form the building [21].

This embodied energy constitutes the initial energy value associated with the building, similar to the economical concept of initial capital or investment.

Considering CO_2 emissions in a conventional residential building with an estimated useful life of 50 years, this initial investment in materials normally constitutes 30% of total emissions. In buildings with high energy efficiency, the energy consumed during the life cycle of the building can drop to even less than 70%. In the latest, the embodied energy percentage increases significantly with respect to the total energy consumption (in some cases this reaches 50%).

The study performed with REAP software to analyse the EF of standard housing in London, considers embodied energy separately from its transport [22]. The results in terms of EF are summarized in Table **29**. The EF of material consumption in the construction of dwellings is 0.11 gha/cap, where the embodied energy concept does not include the associated transport, and this transport, of 0.0099 gha/cap, is not negligible if compared to the incorporated energy values.

Table 29. EF of construction materials [22].

Components	EF (gha/cap)
Embodied energy	0.0687
Transport (of materials)	0.0099
Land (directly used)	0.0364
Total	0.1151

Two main international databases for construction materials have been identified, Eco Ecoinvent [23] and ELCD core [24]. Ecoinvent was developed by the Swiss Centre for Life Cycle Inventories. Due to its consistency and transparency, it has been included in SimaPro LCA software. Ecoinvent is perfectly suited for construction purposes, since every construction material category is included and developed with a high variety of products. The second database is the ELCD core database version II which is supported by the European Commission. The ELCD comprises more than 300 entries including some key materials, transport, and waste management systems [24]. It was last updated in 2005. The database is accessible free of charge, and it is also included in SimaPro and in GaBi LCA software.

The data sets for construction materials can be complemented with other databases such as GaBi Database and PlasticsEurope Eco-Profiles. GaBi Database is one of the biggest LCA databases in the current market [25]. More than 1000 processes for construction materials are included, many of which are from PlasticsEurope, ELCD or Eurofer, and are predominantly cradle-to-gate processes. This is a complete database that refers to construction materials, with sufficient variety within each category of products.

Eco-Profiles were developed by PlasticsEurope in 1991, but they have been continuously updated. This is a free LCA database specializing in plastic materials [26]. Cradle-to-gate data for major polymers as they are produced in Europe is provided. This database is included in SimaPro under the name of Industry data v2.0, and in GaBi as PlasticsEurope.

In Spain, another source of information is ITEC data (Institute of Construction Tech-nology of Catalonia) [27], which shows the values of *primary energy or embodied energy content*, expressed in MJ/kg, of building materials [28]. These values have been obtained from various sources and refer to the energy content of the materials in the processes of raw material extraction, manufacture, transformation, associated transport, installation, maintenance and disposal, without considering, for example, the corresponding part of the energy to be spent on the construction and maintenance of infrastructures or on specific resources for their execution. The human energy used in manpower processes is also excluded.

The TCQ 2000 software [27], developed by ITEC, manages the technical, economical and temporal data involved in the construction work in an integrated way, and includes environmental management. This allows the environmental impact of construction materials to be analysed, such as:

- Energy consumption in the manufacturing of materials
- CO_2 emissions produced by manufacturing
- Energy consumption in the installation of materials
- CO_2 emissions produced by the installation

thereby obtaining:

- Energy cost of materials (energy consumption in the manufacture of materials that are part of the project budget)
- CO_2 emissions of materials

The results of this software tool include the energy cost of the materials, expressed as energy consumption during material manufacturing. The energy consumption includes the processes associated with raw material extraction, manufacturing, processing, transport, installation, maintenance and disposal. The inclusion of transport within the embodied energy concept assumes that the energy spent on transportation to the construction site is small compared to the total; otherwise, the transportation analysis must be separated from the embodied energy of materials.

The embodied energy data reliability, referred to as the energy consumed by a specified material, is closely related to the possibilities of accessing accurate information sources, frequently unavailable, and to the considerations of variations based on the scope of the application (local, regional, national and international). Depending on the sources, the definitions remain very similar but not identical. The ITEC data is henceforth taken as accurate, although contrasted. The other sources have been discarded due to the major disparity in their results [28]. Appendix **D** summarizes representative values of the specific embodied energy (MJ/kg) of the most relevant building materials, according to various sources, and significant differences can be observed.

Source 1 [28]: General Directorate of Housing, Architecture and Urbanism, Cerdá Institute, IDAE, 1999. Guide to Sustainable Building, Energy and Environmental Quality in Construction. Ministry of Development. Madrid, Spain.

Source 2 [12]: Cuchí i Burgos A., López Caballero I., 1999. Informe MIES: una Aproximación al Impacto Ambiental de la Escuela de Arquitectura del Vallés (MIES Report: An Approach to the Environmental Impact of the School of Architecture of the Vallés). Polytechnic University of Catalonia (UPC). Barcelona, Spain.

Source 3 [29]: Cuchí i Burgos A., 2005. Arquitectura i Sostenibilitat. (Architecture and Sustainability) Polytechnic University of Catalonia (UPC). Barcelona, Spain.

Source 4 [27]: ITEC, 2005. Metabase-TCQ 2000: Environmental data. ITEC. Barcelona, Spain.

Source 5 [22]: Nye M., Rydin Y., 2008. The Contribution of Ecological Footprinting to Planning Policy Development: Using REAP to Evaluate Policies for Sustainable Housing Construction. Environment and Planning B: Planning and Design 35(2) 227 – 247. This data does not include transport, which was estimated to be within a range of 10 to 150 km.

Source 6 [19]: Berge B., 2009. The Ecology of Building Materials. Architectural Press. Amsterdam, Netherlands. This data does not include transport. In addition, the combustion value of the product is included in the embodied energy concept, which especially influences plastic materials and those made of wood.

The values of data sources 1 to 4 are similar, as all of them come from ITEC studies, but there are minor variations in some materials due to the year of publication. The results from sources 5 and 6 differ, but are generally approximate. The most appreciable differences are found in those materials which have a significant combustion power, such as wood and plastics. In the case of source 6, the value of combustion is generally included for these materials, whereas in ITEC sources, the value is not included. Furthermore, the material transport to the construction site is not included in cases 5 and 6, which slightly varies the values.

From these figures, the material which has the highest energy impact is that of aluminium, although its low weight greatly reduces its impact on the total embodied energy of the building. In addition, it has a high level of recyclability, thereby allowing it to be recycled many times without loss of quality, and at a low cost to the environment. In Spain, aluminium recycling rate is 30% [28].

Other metals also incorporate significant amounts of energy in their manufacturing processes, but have a high recycling capacity which requires lower amounts of energy. The most important of these metals is steel, which, although it has lower energy consumption per kg than aluminium, it has a greater total impact due to the large amounts consumed in building construction. In fact, its energy impact is usually highly relevant in Spanish constructions. The recycling rate is 20% and its useful life generally exceeds that of the building. Another construction material with high embodied energy is plastic, although its low weight makes its global impact low.

Another construction material, reinforced concrete, is directly responsible for CO_2 emissions, and steel reinforcement and cement are the two most significant sources of emissions, due to their dominant presence in current construction. One alternative for the reduction of its impact is the use of substitute materials, such as ceramic masonry, which is a material widely used in construction, with an impact similar to that of concrete, but with greater possibilities for environmental improvement.

One way to reduce the embodied energy in construction would be through the introduction of organic materials, such as wood and its derivatives, which are normally only used in secondary applications (carpentry and pavements). Another action to reduce embodied energy involves the use of recycled materials in all possible applications, especially recycled aggregates and metals.

Industrialization processes, which increase the life cycle of non-renewable materials and reduce the waste volume, can be incorporated into the construction in order to minimize the energy consumption. The use of dry construction systems also enables execution time and material losses to be minimized on site. With actions such as those discussed above, emissions can be reduced by up to 30% during the building construction [22].

The EF of Building Materials

In the previous section, the concept of embodied energy and how that energy can be quantified have been analysed. The next step is to evaluate the EF of these materials, and the applicability to the present methodology.

The first reference included is that to Chambers research [1]; the data is summarized in Table **30**. The number of materials for which footprint values are obtained is reduced and the variability of the footprint values is high, possibly due to the small amount of information on embodied energy available. Each of these footprints includes the embodied energy and the land use, as in the case of steel, where the percentage of energy used in the extraction from the quarry is considered, or, in the case of cement, the embodied energy and the land used by the factories and the energy in its construction are also taken into account. However, in the present methodology, pre-calculated footprint values are not employed but, instead, the material embodied energy data is used to determine the footprint values.

The second reference is REAP, which presents the result of the EF study of standard apartment buildings in London [22]. REAP analysis by component gives a material footprint of 0.11 gha/cap.

Table 30. EF of construction materials.

Materials	EF (ha/year per tonne)
Wood	1.00 (soft) - 5.70 (hard)
Cement	0.10
Steel	0.80 - 1.40
Paper	2.80 - 4.00
Glass	1.00 - 1.10
Plastic	3.60 - 4.10

The last reference is Domenech [3] who, for the calculation of the footprint associated with consumption of building materials, employs the polynomial formulae [30] defined by the Spanish Ministry of Development. The formulae assign a percentage of different concepts that make up a construction work (manpower, energy, cement, steel materials, bituminous binders, ceramics, wood, copper and aluminium). The coefficients obtained give a reference of the importance that each material commands within the project. However, this procedure does not form the basis of the present calculation, although it does provide a means of comparison. The list of materials in the polynomial formulae fails to include all possible materials which are part of a construction project;

certain information for the calculation of the footprint is excluded. Moreover, the weights assigned to materials fail to consider the constructive building typology. In addition, the percentages of polynomial formulas are weighted on the project budget, and not on the actual weight of the work carried out. The information concerning the embodied energy of the materials is available and can be compared to the results of the polynomial formulae.

Therefore, the procedure for determining the EF of construction materials is as follows (Fig. **15**):

1. Application of embodied energy values listed in the Appendix D. An average of the available values is used, provided that there is no major difference between values, otherwise the outlying values are discarded. The embodied energy includes manufacture, transport and installation of materials.
2. Determination of the consumption (by weight) of each material from the project budget. Polynomial formulae must also be taken into account if any information on material consumption remains unavailable.

$$Eem_i = Cm_i.Esem_i \qquad \qquad \ldots \qquad \qquad \textbf{(23)}$$

where:
Eem_i: Embodied energy (MJ)
Cm_i: material consumption i (kg)
$Esem_i$: Specific-material embodied energy i (MJ/kg)

3. Determination of the footprint (energy) of construction materials. It is assumed that the manufacturing processes of materials are energy-intensive, and that this energy comes from fossil fuels, usually oil. The energy productivity of oil is 71 GJ/ha. Therefore, determining the footprint of construction materials is performed using the following expression:

$$EF_m = \frac{\sum_i Cm_i \cdot Esem_i}{EP} \qquad \qquad \ldots \qquad \qquad \textbf{(24)}$$

where:

EF_m: EF of building materials (ha)
Cm_i: material consumption (kg)
$Esme_i$: Specific-material embodied energy (MJ/kg)
EP: Oil energy productivity (MJ/ha). This value is 71,000 MJ/ha.

This productive land is the total footprint of construction materials, assuming that the production energy of all the manufacturing processes is obtained from oil. This productive land also includes consumption associated to materials that are part of the indirect costs, such as auxiliary installations.

The energy consumed in the manufacture of the construction machinery used on the site is not part of this footprint, since it is considered part of the footprint of the machinery manufacturers, and not of the footprint of the construction site. The allocation of an extra energy cost, due to machine depreciation, is included in the fuel consumption of machinery (10% increased), in the same way as carried out with cars in the mobility section.

CONCLUDING REMARKS

In the proposed analysis, manpower and construction materials are considered impact sources that consume resources indirectly, that is, the impact is caused not by the source, but by its components. First, the manpower consumption is studied by focusing on the most determinant aspects of its impact: food and mobility. The transformation of these types of consumption into EF values is performed by previously documented processes which are adapted to the specific characteristics of the building sector. Data from international statistics, such as FAO and GFN, are used to determine the food consumed per construction worker and the productivities associated to each food family identified, respectively.

The EF associated to the consumption of construction materials during the building execution process takes into account the energy consumption deriving from the manufacture and transport of each of the materials used in the construction of buildings. The construction materials and products present an additional difficulty in its EF analysis with respect to other building construction elements. This is due to the lack of trans-parency in many of the LCA databases consulted, due to the absence of documentation and references in certain cases. It has been detected that databases tend to merge in order to generate more complete data sets, which is totally positive for the development of databases with a greater territory scope, as well as for an increased capacity in categories and content of materials. In a further analysis, it has been detected that results noticeably fluctuate depending on the database used. It does not seem possible that the manufacturing characteristics for the different countries where the LCA studies have been carried out may vary that much, which calls for further research into the sources.

Evaluation of the environmental impact caused by construction materials frequently presents such obstacles as the mismatch between the construction project location and where the LCA database was made, lack of transparency, and/or the unsuitability of the data to the building project conditions. In the present evaluation, 6 sources are analysed and an average embodied energy is used.

CONFLICT OF INTEREST

The author confirms that this chapter has no conflict of interest.

ACKNOWLEDGEMENT

Ministry of Innovation and Science, through the concession of the R+D+I project: Evaluation of the EF of construction in the residential sector in Spain. (EVAHLED). 2012-2014. *Ministerio de Innovación y Ciencia, por la concesión del Proyecto I+D+i: Evaluación de la huella ecológica de la edificación en el sector residencial en España (EVAHLED). 2012-2014.*

REFERENCES

[1]   N. Chambers, C. Simmons, and M. Wackernagel, *Sharing Nature's Interest: Ecological Footprints as an Indicator of Sustainability.* Sterling Earthscan: London, Great Britain, 2004.

[2]   M. Marrero, A. Freire-Guerrero, J. Solís-Guzmán, and C. Rivero-Camacho, "Estudio de la huella ecológica de la transformación del uso del suelo", *Seguridad y Medio Ambiente,* vol. 136, pp. 6-14, 2014. ISSN: 1888-5438. http://www.mapfre.com/documentacion/publico/i18n/catalogo_imagenes/grupo.cmd?path=1081726. [Accessed Feb 13, 2015]

[3]   J.L. Domenech Quesada, *Huella ecológica y desarrollo sostenible (Ecological Footprint and Sustainable Development).* AENOR: Madrid, Spain, 2007.

[4]   FAOSTAT, *Statistic division of the Food and Agriculture Organization of the United Nations,* 2014. http://faostat3.fao.org/home/E [Accessed 21st December 2014].

[5]   *National Footprint Accounts Workbook Learning License,* 2014. http://www.footprintnetwork.org/en/ index.php/GFN/page/national_footprint_accounts_license_academic_edition/ [Accessed 21st December 2014].

[6]   J. Solís-Guzmán, M. Marrero, and A. Ramírez-de-Arellano, "Methodology for determining the ecological footprint of the construction of residential buildings in Andalusia (Spain)", *Ecol. Indic.,* vol. 25, pp. 239-249, 2013.
[http://dx.doi.org/10.1016/j.ecolind.2012.10.008]

[7]   P. González-Vallejo, M. Marrero, and J. Solís-Guzmán, "The ecological footprint of dwelling construction in Spain", *Ecol. Indic.,* vol. 52, pp. 75-84, 2015.
[http://dx.doi.org/10.1016/j.ecolind.2014.11.016]

[8]   M. Wackernagel, and W. Rees, "Our Ecological Footprint: Reducing Human Impact on the Earth. British Columbia, Gabriola Island", *New Soc.,* 1996.

[9]   Instituto para la Diversificación y Ahorro de la Energía (IDAE), *Guía Práctica de la Energía: Consumo Eficiente y Responsable (Practical Energy Guide: Efficient and Responsible Consumption).* IDEA: Madrid, Spain, 2007. [Online] Available: http://www.idae.es/index.php/mod.documentos/ mem.descarga?file=/documentos_11406_Guia_Practica_Energia_3ed_A2010_509f8287.pdf. [Accessed Sept 18, 2014].

[10]   M. Calvo, *Ordenación del Territorio, Urbanismo y Movilidad desde un Enfoque de Huella Ecológica, Seminario: La Huella Ecológica en España (Spatial Planning, Urban and Mobility from an ecological footprint approach. Seminar: The Ecological Footprint in Spain),* October 22-23, 2007. Fundación Biodiversidad: Madrid, Spain.

[11]   M.E. Figueroa Clemente, and S. Redondo Gómez, *Los Sumideros Naturales de CO$_2$: una Estrategia Sostenible entre el Cambio Climático y el Protocolo de Kyoto. (Natural sinks of CO$_2$: A Sustainable Strategy between Climate Change and the Kyoto Protocol).* Universidad de Sevilla: Seville, Spain, 2007.

[12]   A. Cuchí, and I. López Caballero, *Informe MIES (MIES Report: An Approach to Environmental Impact of the School of Architecture of the Vallès).* Universidad Politécnica de Cataluña (UPC): Barcelona, Spain, 1999.

[13]   Best Foot Forward (BFF), *City Limits: A Resource Flow and Ecological Footprint Analysis for Greater London.* Chartered Institute of Wastes Management Body Environment: London, Great Britain, 2002.

[14]   J. Barrett, "Component ecological footprint: sustainable developing scenarios", *Impact Assess. and Appraisal,* vol. 19, pp. 107-118, 2001.
[http://dx.doi.org/10.3152/147154601781767069]

[15]   Going Green Limited, *ECOCAL. Version 4.2.3.* Best Foot Forward, 2005 [Online] Available: www.bestfootforward.com. [Accessed Sept 10, 2014].

[16]   Carbon Rationing Action Groups (CRAG), *CO$_2$ Conversion Spreadsheet: Calculate your Carbon Emissions.* CRAG, 2007. [Online] Available: http://www.carbonrationing.org.uk. [Accessed Sept 10,

2014].

[17]   AEA Energy & Environment, *Tool for Calculations of CO₂ Emissions from Organisations.* Department for Environment, Food and Rural Affairs (DEFRA): Great Britain, 2008.

[18]   I. Arto, and D. Pon, *Escenarios de Evolución de la Huella Ecológica.*, Seminario: La Huella Ecológica en España (Evolution Scenarios Footprint. Seminar: The Ecological Footprint in Spain), October 22-23, 2007. Fundación Biodiversidad: Madrid, Spain.

[19]   B. Berge, *The Ecology of Building Materials.* Architectural Press: Amsterdam, Holland, 2009.

[20]   G. Wadel, J. Avellaneda, and A. Cuchí, "La sostenibilidad en la arquitectura industrializada: cerrando el ciclo de los materiales. (Sustainability in architecture industrialized: closing the materials cycle)", *Informes de la Construcción,* vol. 62, no. 517, pp. 37-51, 2010. [http://dx.doi.org/10.3989/ic.09.067]

[21]   T. Solanas, and J. Herreros, *Vivienda y sostenibilidad en España. Volume 2: Colectiva (Housing and Sustainability in Spain. Volume 2: Collective).* Gustavo Gili: Barcelona, Spain, 2008.

[22]   M. Nye, and Y. Rydin, "The contribution of ecological footprinting to planning policy development: using REAP to evaluate housing policies for sustainable construction", *Environ. Plann. B Plann. Des.,* vol. 35, no. 2, pp. 227-247, 2008. [http://dx.doi.org/10.1068/b3379]

[23]   Ecoinvent Centre, *Ecoinvent website,* 2013. http://www.ecoinvent.org/database/ [Accessed Jan 3, 2013].

[24]   ELCD, *ELCD core database II website,* 2013. http://lca.jrc.ec.europa.eu/lcainfohub/datasetArea.vm [Accessed Jan 3, 2013].

[25]   PE INTERNATIONAL, *GaBi Database website,* 2013. Construction Materials extension. http://www.gabi-software.com/support/gabi/abi-lci-documentation/data-sets-by-database-modules/ extension-databases/xiv-construction-materials [Accessed Jan 3, 2013].

[26]   Plastics Europe, *Plastics Europe Eco-Profiles website,* 2013. Construction Materials extension. http://www.plasticseurope.org/plastics-sustainability/eco-profiles.aspx [Accessed Jan 3, 2013].

[27]   ITEC, *Metabase-TCQ 2000: Datos Ambientales (Environmental Data),* ITEC: Barcelona, Spain, 2005. [Online] Available: http://www.itec.es/programas/tcq/. [Accessed Sept 15, 2011].

[28]   General Directorate of Housing, Architecture and Urbanism, Cerdá Institute, and IDAE, *Architecture and Urbanism, Cerdá Institute, and IDAE, Guide to Sustainable Building, Energy and Environmental Quality in Construction.* Ministry of Development: Madrid, Spain, 1999.

[29]   A. Cuchí, *Arquitectura y sostenibilidad (Architecture and Sustainability).* Universidad Politécnica de Cataluña (UPC): Barcelona, Spain, 2005.

[30]   Spain MP (Ministry of the Presidency), Real Decreto 3650/1970, de 19 de Diciembre, por el que se aprueba el cuadro de fórmulas-tipo generales de revisión de precios de los contratos de obras del Estado y Organismos autónomos para el año 1971 (Royal Decree 3650/1970, of December 19, through which is approved the set of general formulae of the revision of costs for works contracts of the State and autonomous bodies for the year 1971). Madrid, Spain, 1970.

CHAPTER 5

Waste and the Constructed Area

Abstract: In this chapter, the environmental impact of waste and the constructed area are analysed. The waste is defined as those residues most relevant to the present model: urban waste and construction and demolition waste. For the urban waste, the generation estimates per person per year from statistical data are employed. In the case of the CDW, generation estimates come from a software tool, developed, among others, by the present authors and, which gives, according to the residential typology considered, the CDW volume expected.

Once the expected waste volumes are determined, the waste analysis is based on the methodology found in Wackernagel´s studies into the determination of its footprint. His work establishes that the footprint associated with waste disposal, emissions, and/or discharges is calculated in the same way as for the materials: the same energy intensity (embodied energy) is applied but the percentage of energy that can be recovered for recycling is deducted.

In the constructed land EF calculation, only the land used for urbanization and buildings is considered. In this case, a conversion factor is unnecessary because the units are already in terms of surface area, and the area passes from m^2 to ha. The equivalence factor is that of agricultural land, since most of the infrastructure and built land are located in areas of agricultural quality.

Keywords: Absorption factor, Construction and demolition waste, Conversion factors, Dwelling construction, Ecological footprint, Embodied energy, Emission factor, Natural productivity, Productive land, Urban solid waste.

WASTE

In this section, the environmental impact of waste is analysed. This analysis focuses on those residues most relevant to the present model: urban waste and construction and demolition waste (CDW). The waste generated throughout the life cycle of the building are varied and of diverse origins. Focusing on the building construction phase, two main classes can be identified: municipal solid waste (MSW), generated on the construction site; and the CDW generated by the

Jaime Solis-Guzman and Madelyn Marrero

construction materials and their packaging. The municipal solid waste is distributed as in Fig. (**18**) [1] and the CDW in Fig. (**19**) [1].

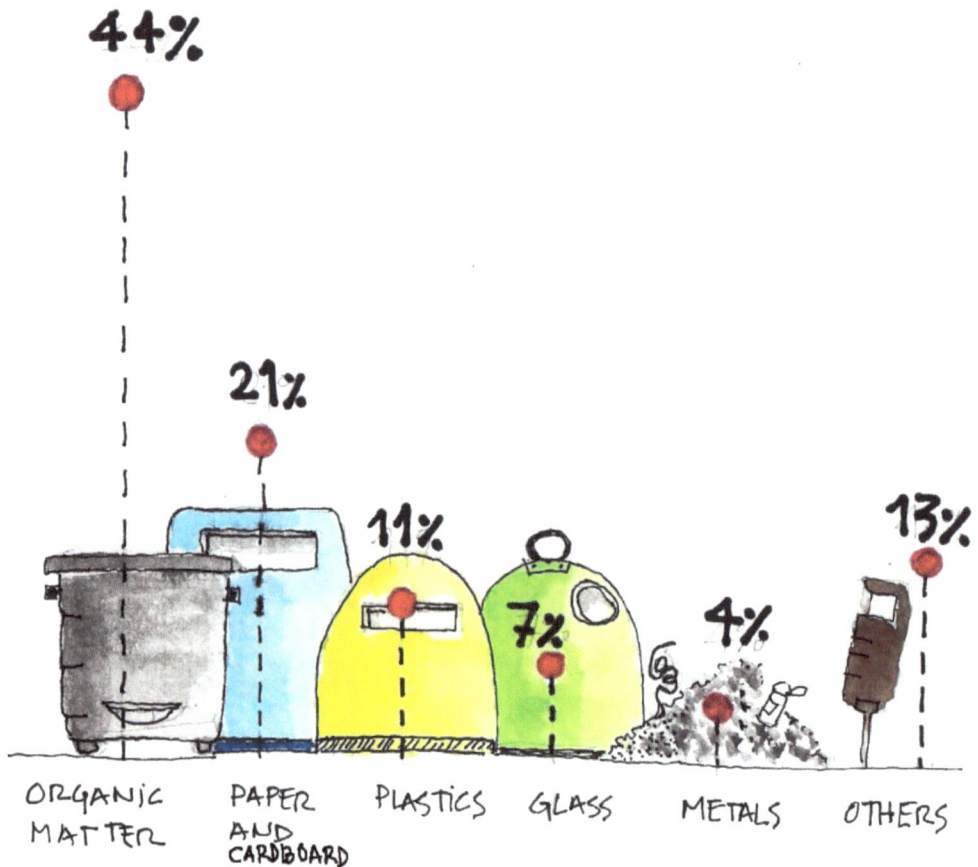

Fig. (18). The percentages of the MSW components.

Fig. (**20**) shows the methodology for calculating the waste EF to be followed in this section. In the same figure, the top level, or first step, is to determine the type and quantities of the waste generated. That is, the volume of waste that is produced on the construction site. There are the two main categories: municipal solid waste (MSW) and construction and demolition waste (CDW).

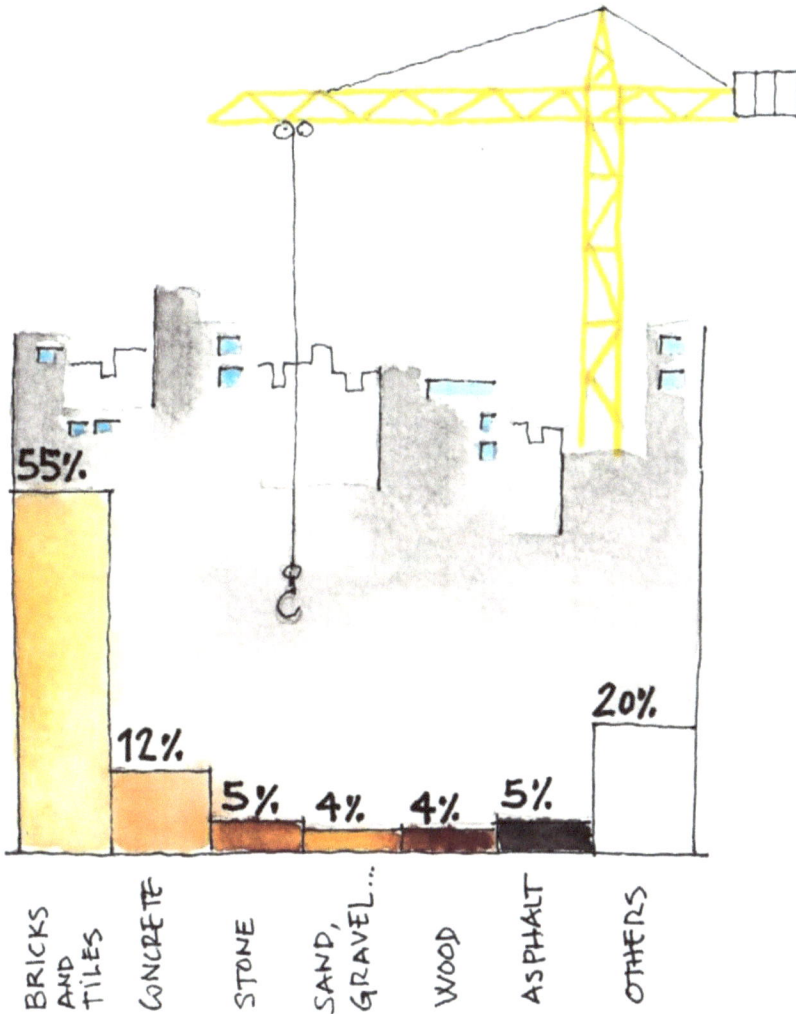

Fig. (19). The percentages of the CDW components.

In 2007, generation estimates are of 516 kg per person per year or 1.41 kg per person per day [2]. These amounts refer to household urban waste, with or without selective collection and without considering those from construction and demolition, or urban waste from industry.

The CDW, mainly generated during the construction and demolition phase, is waste, which, due to its high volume, needs to be managed properly. Spain generates 600 to 1,000 kg of CDW per habitant per year, making this type of waste a significant quantity in the analysis of the building construction footprint [3]. The percentages of the components within the CDW [1] are represented in Fig. (**19**).

Many models have been established over the last decade to determine the project waste quantities, such as OEKO Service Luxembourg [4], which proposes quantification of C&DW at the worksite, and is able to estimate types and volumes produced. The National Technical University of Athens (NTUA) has developed an indicative mathematical model for the estimation of the generated quantities of C&DW [5]. SMARTWaste™ is another quantification method, applied in the United Kingdom, and is based on data obtained from previous experiences and calculates the waste volumes in 13 categories: ceramic, concrete, wooden pallet, *etc*. [6]. More recently, a review on the various quantification methods has been published [7].

The present authors, together with others, have also developed a quantification model to estimate the type and quantity of waste generated by different construction projects, such as new buildings, demolition, renovations and alterations [8]. This software is free. Over recent years, the model has been tested at the Los Alcores (Seville, Spain) Community treatment plants. The classification code used is the same as that which Spanish quantity surveyors normally employ to obtain the bill of quantities, thereby making the model both easy to understand and to implement [9]. More recently the model has been adapted to road construction [10]. The software tool is free and following information is required in order to perform the calculations:

1. Project type (demolition or new construction)
2. Foundation type (concrete slab, pile, trenches, or concrete pads)
3. Number of storeys (with basement and regardless of business premises on the ground level)
4. Constructed surface in square metres.
5. Results: total volume of earth in cubic metres, mixed CDW in cubic metres, and the cost of the municipal licence.

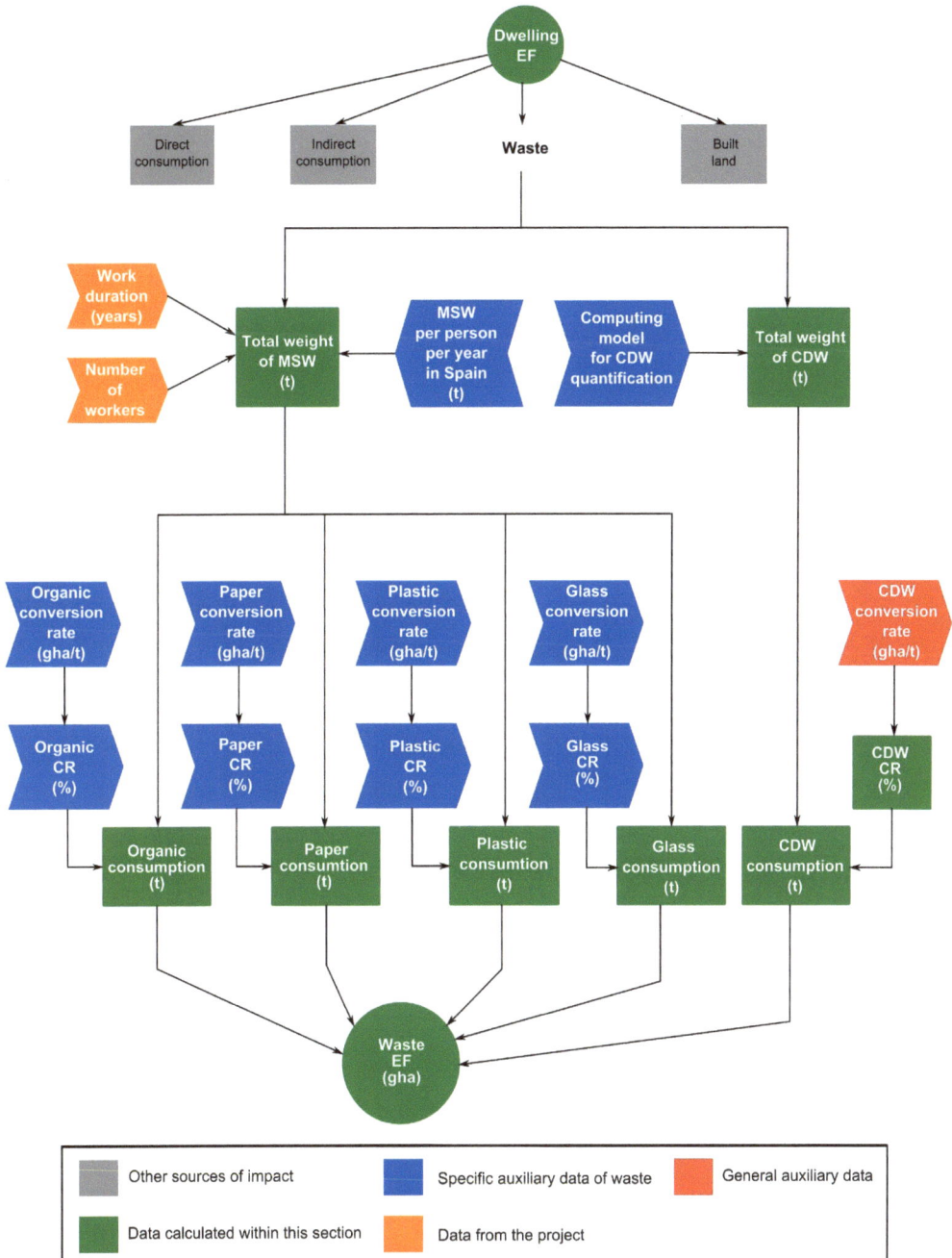

Fig. (20). Method for determination of the waste EF.

Once the waste volumes are obtained from the software, the next step is to define the waste recycling rates: higher rates generate lower energy impact.

The concept of recycling rate can be defined as follows [11]:

$$Recycling\ rate\ (kg/kg)\ =\ \frac{recycling\ (kg)}{stock\ (kg)}\ ...\qquad (25)$$

That is, the weight of material that is recycled with respect to the total production.

MSW recycling rates are represented in Table **31**. For the case of organic waste, the national data is used, which determines the percentage treated for composting [12]. The same source provides information on the recycling rate of organic waste with respect to the total urban waste, around 30%. However, this rate may incorporate treatments that exclude the effective recovery of the organic waste. For the other flows (paper, plastics, and glass), the data on the recycling rates from the Department of Environment of Andalusia [2] is used.

The CDW types generated on the construction site are not separated and arrive to the treatment plants in a mixed state. Indeed, although the legislation in Spain [8] requires CDW separation at origin, the construction activity in Spain has yet to establish the processes. In the case that this separation takes place; the calculation of the various fractions [13] is needed in order to evaluate the corresponding footprint.

Table 31. Recycling rates of MSW.

Fractions	Recycling (%)
Organic	12-15
Paper and cardboard	50
Plastics	40
Glass	40

A recycling rate of 15% [3], an estimate well below the national and European objectives, is used in the present analysis. The Spanish objective was 40% in 2011 [14] and that of the European Union is 70% in 2020 [15]. In Spain, the recycling rates are still far from those of leading countries, such as the Netherlands and Germany.

Actions which reduce the consumption of resources and waste generation include that of the reduction in the demand for non-renewable resources by means of the reuse and recycling, not only of building materials, but of the buildings themselves and their urban environment [16].

The present analysis is based on the methodology found in Wackernagel's studies into the determination of the waste footprint [17]. His work establishes that the footprint associated with waste disposal, emissions, and/or discharges is calculated in the same way as for the materials: the same energy intensity is applied but the percentage of energy that can be recovered for recycling is deducted.

According to Wackernagel, it is estimated that recycling paper and cardboard can recover 50% of the embodied energy. That is, if, for example, the footprint of x tonnes of paper waste is 20 ha, with 100% recycling, the footprint would be 10 ha. The estimates are summarized in Table **32**.

Table 32. Energy recovery due to recycling.

Material	Energy recovery per recycling (%)
Organic waste	100 (compositing)
MSW	50
Plastics	70
Glass	40
Paper and cardboard	50
Aluminum	90
Magnetic metals	50
Debris	90

Paper waste has a fossil footprint (CO_2 emissions associated with energy consumption in the production processes) and also a forest consumption footprint for its raw material when not obtained from recycled paper. There are studies which already assess the waste footprint. In Chambers' studies [18], data of the footprint of recycled and non-recycled materials is collected, thereby providing evidence of the difference in impacts. For example, the paper footprint was estimated in the range of 2.8-4 ha/year per tonne, while the recycled-paper footprint showed a result of 2-2.9 ha/year per tonne. The difference in footprint is due to the fact that recycled paper consumes no forest land in its manufacture and the energy consumption in manufacturing is much lower, at about 30%. Similar results are found for the analyses of glass and of plastics.

Another calculation of the waste footprint can be found in REAP software, which analyses London dwellings, where the CDW recycling rate is 25% and considers that the material embodied energy is decreased when recycled [19].

In general, when materials are recycled, multiple savings take place since recycled materials require less energy in their processing and reduce material extractions from nature, and can also lead to a proportional reduction in landfill.

In the present methodology, all consumptions are charged to the fossil footprint. Therefore, this methodology excludes the built land footprint from landfills and treatment plants, since that footprint is included in the regional or local analysis, and cannot therefore be considered as part of the impact of building construction processes. There are other approaches, however, which include the built land footprint from landfills and treatment plants [20].

The methodology here is based on Domenech's research [17] and is expressed in recycling rates. These rates can refer to very different waste origins (such as hazardous, non-hazardous, and paper). In the present case study, two kinds of waste are defined: non-hazardous and paper origin.

For *non-hazardous waste*, the energy intensity of the production of the material of which the waste is composed is used, and the percentage of energy that can be recovered by recycling is discounted [20]. Some of these non-hazardous wastes are organic or CDW. The calculation is as follows:

$$CR_x = \frac{EI_x}{EP} \cdot (1 - \frac{\%R_x}{100} \cdot \frac{\%SE_x}{100}) \quad \ldots \qquad \text{(26)}$$

where:

CR_x: conversion rate of non-hazardous waste x (ha/t).

EI_x: energy intensity of material production. The energy intensities of the production of the materials to be recycled are summarized in Table **33**.

EP: energy productivity of waste (which is assumed equal to that of fossil fuels). It is therefore taken as 71 GJ/ha per year.

% R_x: recycling percentage. The urban waste values from Table **31** are used and 15% for CDW.

% SE_x: percentage of energy recovered by recycling, (Table **32**).

Table 33. Energy intensity of consumer goods.

Consumer goods	EI (GJ/t) [17]	EI (GJ/t) [21]
Raw mineral	1.50	
CDW (debris)	5.00	
Manufacture of cement, plaster, stone, soil	5.00	
Organics	20	
Glass	20	
Paper and cardboard	30	35
Basic products of iron, steel, and other metals	30	50
Plastic derivatives	43.75	50

In order to express the conversion rate in gha/t, the equivalence factor of fossil energy FE_F is applied, and the equation becomes:

$$CR_x = \frac{EI_x}{EP} \cdot (1 - \frac{\%R_x}{100} \cdot \frac{\%SE_x}{100}) \cdot e_f \quad \ldots \qquad (27)$$

where:

CR_x: weighted conversion rate (gha/t)

For paper and cardboard, the forest footprint is added to the fossil energy footprint [20]:

$$CR_x = \frac{EI_x}{EP} \cdot (1 - \frac{\%R_x}{100} \cdot \frac{\%SE_x}{100}) \cdot e_f + \frac{1}{NP} \cdot (1 - \frac{\%R_x}{100} \cdot 0.8) \cdot e_f \quad \ldots \quad (28)$$

where:

NP: paper natural productivity. This is considered 1.01 t/ha per year [17].

% R_x: recycling rate, 50%.

0.8 is a correction factor analogous to the energy recovered by recycling of the previous equation, which represents the productivity that can be restored with recycled paper (about 80%).

Finally, the equation to determine the total waste footprint is:

$$EF_{wws} = \sum_i CR_{x_i} \cdot G_i \qquad \ldots \qquad \textbf{(29)}$$

where:

EF_{wws}: Weighted EF of waste (gha)

CR_{xi}: Weighted conversion rate (gha/t)

G_i: Waste generation (t)

In conclusion, the procedure in the methodology defined in Fig. (**20**) can be summarised as the following:

1. Determination of MSW and CDW volumes or weights generated on the construction site. There are five waste types: organic, paper, plastics, glass, and CDW. These cal-culations are based on estimates (Figs **18** and **19**) or computer tools.
2. Determination of conversion rates (CR) for each waste type, following the formulation discussed.
3. Calculation of the Waste EF (forest and energy footprint).

The hazardous waste constitutes a major issue in demolition projects [22] but is not that important in new construction. It is not part of the present methodology due to its low impact in the complete analysis: potentially hazardous waste represents less than 2% of the total construction waste volume, and is essentially composed of cans and containers for paints [8]. Another potentially hazardous waste comes from concrete additives; this consumption is even lower: for example, in the case study, additives are 3% with respect to the weight of the paint.

THE CONSTRUCTED AREA

The implantation in a territory changes its ecological value remarkably. The evaluation of this impact source, to quantify the urbanized areas and make it more sustainable, incorporates strategies such as a progressive decrease in new territory consumption, and the recovery of the ecological quality of plots. This requires the development of planning tools that involve aspects outside the building, such as transport, urban sprawl, and natural systems in the territory [21]. The present methodology determines the controlling parameters and reflects these parameters in accordance with the territory.

Fig. (21). Percentage of population and urban area increase (1990-2000).

Fig. (**21**) [23] compares the increases in the urbanized area and population in Spain and Europe during the last decade of the twentieth century. The population increases similarly (close to 5%); however the urbanized area increases in a very different way: 15% in Europe, and 25% in Spain. That is, intensive building land use has taken place in Spain, which means productive and ecological values have been lost.

This analysis of the land directly occupied by buildings is extremely relevant from a sustainable point of view, since land is biologically unproductive from the moment it is occupied. However, on EF calculations applied to cities or regions, the load capacity, or biocapacity, includes built land since it is a place where people actually live, although the land is not ecologically productive. That is, homes, gardens and green spaces are considered ecologically productive areas, although they become unproductive in the urbanized case.

The impact sources include the built, paved and, generally, seriously degraded land. According to this definition, this category includes [21]:

• Built land or that used to host infrastructures and equipment.
• Mining areas.
• Landfills.
• Areas under construction. For example, built land associated with wind energy, which also includes the area occupied by access roads.
• Reservoirs. Water reservoir land is included when the productive land occupied produces hydroelectric power. There is a double counting, to a certain extent, because there is no clear division between the reservoirs used for energy purposes and those used solely for water supply.

In the present analysis, only the land used for urbanization and buildings is considered, and therefore the remaining concepts are disregarded. Before summarizing the calculation methodology, research carried out by other researchers into directly occupied EF is analysed.

The first study focuses on housing in Canada during the 1990s [18], and compares the EF generated by different housing types. One of the parameters studied is that of directly occupied land. The EF (ha/cap) is 1.2 in a standard terraced house, 0.9 in a house in a city and 0.5 in an apartment in a building.

According to later studies in Britain using the REAP software, the impact of the land directly used, for a home built in the early twenty-first century, is 0.0364 gha/cap [19]. Similar studies [24] are performed using five different housing typologies, for which the EF per built area is 0.32 gha/cap in the first three, and 0.29 gha/cap for the last two typologies.

These typologies are:

1. Typical house in the UK
2. Newly built (2002)
3. Housing rated "excellent" by Ecohomes (BREEAM tool)
4. BedZED housing type I
5. BedZED housing type II (incorporates more energy-saving strategies)

Beddington Zero Energy Development (BedZED) is a residential estate close to London, built during the period 2000-2002, which strives to be environmentally friendly. It was designed by the architect Bill Dunster.

In Spain, the EF study performed by Daniel Calatayud and Coque Claret [25], 2008), on the Torre Ferrera house designed by Daniel Calatayud [26], consists of the analysis of a 2-home condominium project. Both the projects above provide consumption data in terms of EF, but fail to provide direct values of EF for the constructed land.

As the figures show, there is considerable disparity in the values, mainly because the units in which the parameter is expressed vary depending on the case studied, and coefficients and methodologies differ significantly.

In the present methodology, the previous figures are not included, although they do establish an EF range. The specific calculation methodology is summarized in Fig. (**22**).

The EF of planning, implementing, and directly occupied land or, in a simpler expression, constructed land, is generated by the transformation undergone by the land under study. This footprint is obtained by calculating the area consumed by urbanization and the building analysed. In this case, a conversion factor remains

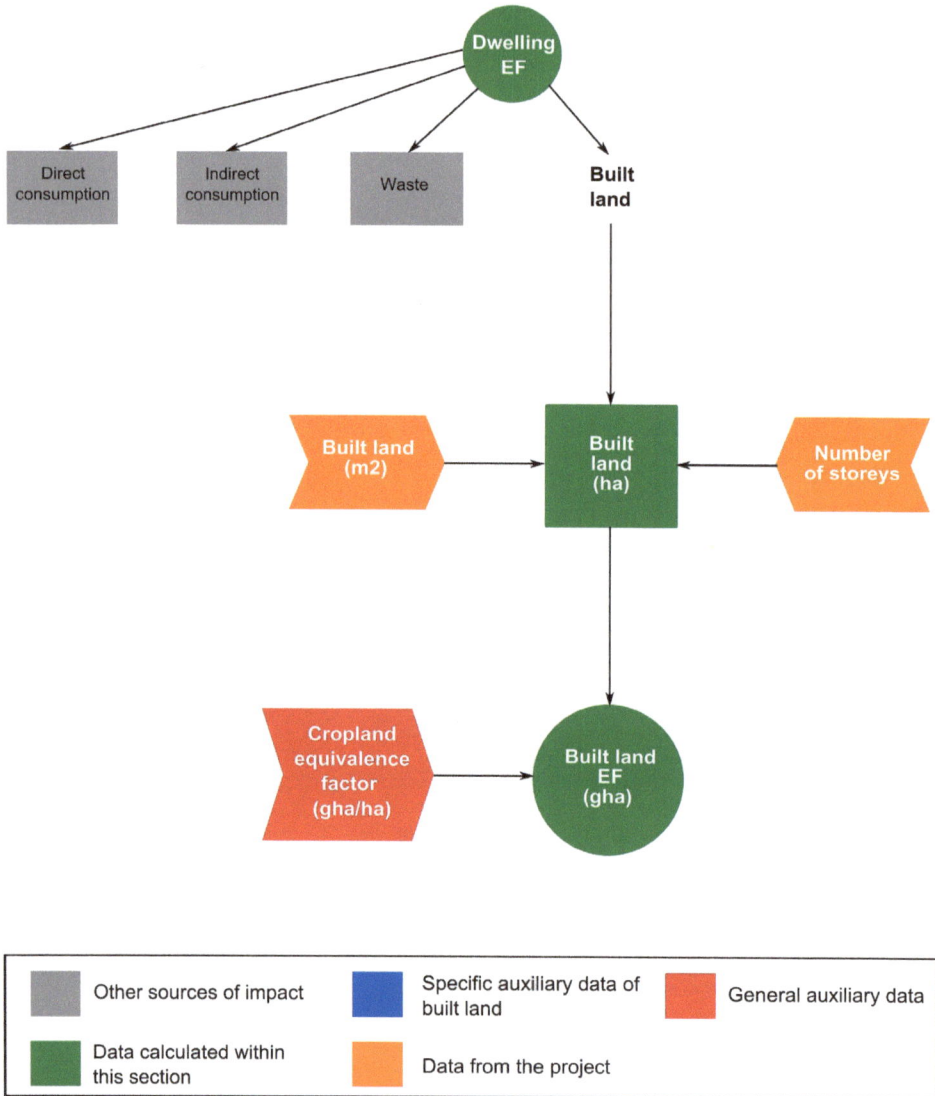

Fig. (22). Methodology for calculating the built land EF.

unnecessary because the units are already in terms of surface area, and the surface passes from m² to ha. As defined by the EF methodology, the area to be computed is given in the form of productive area used directly:

$$EF_b = S \qquad \qquad ... \qquad \qquad \textbf{(30)}$$

where:

EF$_b$: EF of built land (ha)

S: consumed surface (ha)

In order to compare this footprint with the other footprints obtained in the previous sections, the equivalence factor is that of agricultural land, since most of the infrastructure and built land are located in areas of agricultural quality.

Although this statement can be considered to be excessively general, in this case the EF hypotheses and the equivalence factors are not modified.

Table **3** shows the values of equivalence factors. According to this table, the factor for the constructed area (this appears as "settlements" in the table) is the same as that for agricultural land, valued at 2.21 gha/ha.

The EF calculation in global terms therefore becomes:

$$EF_{wb}=S.e_b \qquad \dots \qquad\qquad (31)$$

where:

EF$_{wb}$: Weighted EF of built land (gha)

E$_b$: Equivalence factor of built land

For the calculation of the ecological deficit, the productivity factors are used, and its analysis is performed in the case study.

CONCLUDING REMARKS

The environmental impact of waste and of the constructed area has been analysed. The waste is defined as those residues most relevant to the present model: urban waste and construction and demolition waste. For urban waste, the generation estimates per person per year are employed from statistical data. There are several international tools related to CDW estimation. In the present analysis, a software tool developed by the authors is used to predict the waste quantities.

Once the expected waste volumes are determined, the waste analysis is based on the footprint associated with its disposal, emissions, and/or discharges. This footprint is calculated in the same way as for the materials: the same energy intensity (embodied energy) is applied but the percentage of energy that can be recovered for recycling is deducted.

In the constructed land EF calculation, only the land used for urbanization and buildings is considered. In this case, a conversion factor is unnecessary because the units are already in terms of surface area. The land is always considered to be of agricultural quality as established in the general EF methodology.

CONFLICT OF INTEREST

The author confirms that this chapter has no conflict of interest.

ACKNOWLEDGEMENT

Ministry of Innovation and Science, through the concession of the R+D+I project: Evaluation of the EF of construction in the residential sector in Spain. (EVAHLED). 2012-2014. *Ministerio de Innovación y Ciencia, por la concesión del Proyecto I+D+i: Evaluación de la huella ecológica de la edificación en el sector residencial en España (EVAHLED). 2012-2014.*

REFERENCES

[1]   Spain ME (Ministry of Environment), *(Ministry of Environment), Plan Nacional de Residuos de Construcción y Demolición 2001–2006 (National C&D Waste Plan 2001–2006).* Ministry of the Environment: Madrid, Spain, 2001.

[2]   Andalusia Ministry of Environment, *Environment Report 2008,* Andalusia, Spain, 2009. [Online] Available: http://www.juntadeandalucia.es/medioambiente/site/web/menuitem.318ffa00719ddb10 e89d04650525ea0/?vgnextoid=3b32db0dee134210VgnVCM1000001325e50aRCRD. [Accessed Sept 15, 2011].

[3]   Gremio de Entidades del Reciclaje de Derribos (GERD), *IV Congreso Nacional de Demolición y Reciclaje (IV National Congress of Demolition and Recycling),* May 20-22, 2009, Zaragoza, Spain.

[4]   SuperDrecksKescht fir Betriber, *Oeko-Service Luxembourg. Abfallmanagement im Hochbau mit der Zielvorgabe Abfallvermeidung. LIFE97 ENV/L/000206. Internet Version 31.07.2002 Luxembourg,* 2002. http://www.superdreckskescht.lu [Accessed Sept 15, 2011].

[5]   B. Kourmpanis, A. Papadopoulos, K. Moustakas, M. Stylianou, K.J. Haralambous, and M. Loizidou, "Preliminary study for the management of construction and demolition waste", *Waste Manag. Res.,* vol. 26, no. 3, pp. 267-275, 2008.
[http://dx.doi.org/10.1177/0734242X07083344] [PMID: 18649575]

[6]   SMARTWasteTM, *Building Research Establishment (BRE) Ltd.* Watford, United Kingdom, http:// www.smartwaste.co.uk [Accessed Sept 15, 2011].

[7]   Z. Wu, A.T. Yu, L. Shen, and G. Liu, "Quantifying construction and demolition waste: an analytical review", *Waste Manag.,* vol. 34, no. 9, pp. 1683-1692, 2014.
[http://dx.doi.org/10.1016/j.wasman.2014.05.010] [PMID: 24970618]

[8]   J. Solís-Guzmán, M. Marrero, M.V. Montes-Delgado, and A. Ramírez-de-Arellano, "A Spanish model for quantification and management of construction waste", *Waste Manag.,* vol. 29, no. 9, pp. 2542-2548, 2009.
[http://dx.doi.org/10.1016/j.wasman.2009.05.009] [PMID: 19523801]

[9]   M. Marrero, and A. Ramirez-De-Arellano, "The building cost system in Andalusia: application to construction and demolition waste management", *Construct. Manag. Econ.,* vol. 28, no. 5, pp. 495-507, 2010.
[http://dx.doi.org/10.1080/01446191003735500]

[10]   J. Solís-Guzmán, M. Marrero, and D. Guisado, "Modelo de cuantificación y presupuestación en la gestión de residuos de construcción y demolición. Aplicación a viales. (Model for the quantification and budget of construction and demolition waste management. Application to roads)", *Carreteras,* vol. 195, pp. 6-18, 2014.

[11]   A. Cuchí, *Arquitectura y sostenibilidad (Architecture and Sustainability).* Universidad Politécnica de Cataluña (UPC): Barcelona, Spain, 2005.

[12]   Observatorio de la Sostenibilidad (Observatory of Sustainability in Spain), *Sustainability in Spain 2007,* Madrid, Spain, 2008. [Online] Available: www.sostenibilidad-es.org. [Accessed Sept 15, 2011].

[13]   Spain Ministry of the Presidency, Royal Decree 105/2008, by regulating the production and management of construction and demolition waste, Spain, 2008.

[14]   Spain Ministry of Environment, *II National Waste Plan 2008-2015.* Madrid, Spain, 2008.

[15]   European Parliament, *Waste Framework Directive 2008.* European Union, 2008.

[16]   B. Edwards, *Sustainability Guide.* Gustavo Gili: Barcelona, Spain, 2008.

[17]   J.L. Domenech Quesada, *Huella ecológica y desarrollo sostenible (Ecological Footprint and Sustainable Development).* AENOR: Madrid, Spain, 2007.

[18]   N. Chambers, C. Simmons, and M. Wackernagel, *Sharing Nature's Interest: Ecological Footprints as an Indicator of Sustainability.* Sterling Earthscan: London, Great Britain, 2004.

[19]   M. Nye, and Y. Rydin, "The contribution of ecological footprinting to planning policy development: using REAP to evaluate housing policies for sustainable construction", *Environ. Plann. B Plann. Des.,* vol. 35, no. 2, pp. 227-247, 2008.
[http://dx.doi.org/10.1068/b3379]

[20]   E. Marañón, G. Iregui, J.L. Domenech, Y. Fernández Nava, and M. González, "Propuesta de índices de conversión para la obtención de la huella de los residuos y los vertidos. (Proposed conversion rates to obtain the footprint of waste and effluents)", *Observatorio iberoamericano del desarrollo local y la economía social,* vol. 1, no. 4, April-June 2008.

[21]   G. Acosta Bono, J. González Daimiel, M. Calvo Salazar, and F. Sancho Royo, *Estimación de la Huella Ecológica en Andalucía y Aplicación a la Aglomeración Urbana de Sevilla (Estimation of the Ecological Footprint in Andalusia and Application to the Urban Agglomeration of Seville).* Dirección

General de Ordenación del Territorio y Urbanismo, Consejería de Obras Públicas de la Junta de Andalucía: Seville, Spain, 2001. [Online] Available: http://hdl.handle.net/10326/974 [Accessed Sept 18, 2014].

[22]   M. Marrero, J. Solís Guzmán, B. Molero Alonso, M. Osuna Rodriguez, and A. Ramirez de Arellano, "Demolition waste management in Spanish legislation", *Open Constr. Build. Technol. J.,* vol. 5, suppl. 2-M7, pp. 162-173, 2011.

[23]   Observatorio de la Sostenibilidad (Observatory of Sustainability in Spain), *Sustainability in Spain 2005,* Madrid, Spain, 2006. [Online] Available: http://www.sostenibilidad-es.org. [Accessed Sept 15, 2011].

[24]   T. Wiedmann, J. Barrett, and N. Cherrett, *Sustainability Rating for Homes: The Ecological Footprint Component.* Stockholm Environment Institute: York, Great Britain, 2003. [Online] Available: http://www.sei.se/index.php?section=implement&page=publications. [Accessed Sept 10, 2014].

[25]   T. Solanas, and J. Herreros, *Vivienda y sostenibilidad en España. Volume 2: Colectiva (Housing and Sustainability in Spain. Volume 2: Collective).* Gustavo Gili: Barcelona, Spain, 2008.

[26]   T. Solanas, and J. Herreros, *Vivienda y sostenibilidad en España. Volume 1: Familiar (Housing and Sustainability in Spain. Volume 1: Family).* Gustavo Gili: Barcelona, Spain, 2007.

CHAPTER 6

Case Study

Abstract: A building and urbanization project of one hundred multifamily dwellings in Spain is studied in detail and its ecological footprint (EF) determined. The same methodology is then applied to the construction of other ten projects that include detached, semi-detached and multifamily dwellings. The impact factors are grouped into: direct consumption (energy and water), indirect consumption (manpower and construction materials), waste, and land occupied directly. The manpower impact in building construction is mainly food intake and mobility (workers commuting to the construction site).

For construction material analysis, the project bill of quantities is employed; each material quantity is transformed into its corresponding embodied energy, and expressed in terms of EF. A similar analysis, but using empirical and statistical data, is performed with the power consumption on the construction site and the workers' mobility. The waste generated, which is municipal solid waste and construction and demolition waste, is included in the analysis. Finally, the land directly occupied by the construction project also has a footprint. In summary, each element that forms part of the construction project uses resources (energy, water, manpower, materials) or generates waste, giving rise to an EF. The most important impact in all cases analysed is the embodied energy of construction materials, almost 90%, followed by the food intake by the workforce, 5-9%.

The partial and global footprints obtained are: forest, food, energy, built land, and total EF.

Keywords: Absorption factor, Construction materials, Conversion factors, Dwelling construction, Ecological footprint, Electricity, Emission factor, Forest productivity, Food intake, Fuel, Fuel productivity factor, Productivity factors, Productive land, Standard productive territory, Water consumption, Worker mobility.

INTRODUCTION

In order to validate the methodology proposed in the previous chapters, first, a

Jaime Solis-Guzman and Madelyn Marrero

representative dwelling project in Spain [1, 2] is studied in detail. In second place, the results from other 10 different projects, detached, semi-detached and multifamily dwellings, are analysed and the results compared.

The first project assessed is representative of the most common dwelling type in Spain [1 - 3], which corresponds to 4-storey blocks of flats and commercial offices at ground level, see Fig. (**23**); and the constructed areas per storey and block are listed in Table **34**.

Fig. (23). Representative dwelling building [5].

Table 34. Constructed area per block.

Floor	Block 1 (m²)	Block 2 (m²)
Underground level -2	1,476.09	1,312.08
Underground level -1	1,476.09	1,312.08
Ground floor	1,359.06	1,197.86
First floor	1,359.15	1,197.86
Second floor	1,363.35	1,201.53
Third floor	1,363.35	1,201.53
Roof	113.61	81.28
TOTAL	**8,510.70**	**7,504.22**
Total area constructed	**m²**	**16,014.92**

At the end of the chapter the same methodology is applied to ten dwelling projects of different types and sizes: single family dwellings of 1 and 2 storeys, and multi-family buildings of 3, 4, 5, 6 and 10 storeys, (see Table **35**).

Table 35. Floor area of the ten dwelling projects.

Project number									
1	2	3	4	5	6	7	8	9	10
Number of floors over ground level									
1	2	2	3	3	4	5	5	6	10
Total floor area (m²)									
2697	3836	5754	4440	4440	6661	6662	7772	12211	13320

In all the projects analysed, it is considered that the only activity that takes place on the land is that of the construction activities. This impact lasts one year; the time for the construction to be completed. During the analysis, the electric power and water consumptions are estimated from empirical data of similar building projects. Other impacts are determined from the project bill of quantities and its general characteristics.

METHODOLOGY

The methodology described in previous chapters can be summarized as follows:

1. Defining the impact factors. These are the generators of impact on the land and are classified into: direct consumption, indirect consumption, waste generation and built land. Direct consumption refers to direct use of resources on the construction site, such as energy expenditure (fuel or electricity consumption) and water usage. Indirect consumption is caused by the indirect use of resources, such as material or energy resources from other previous processing:
   ○ Manpower
   ○ Building material consumption

   The manpower during the building construction involves food intake by the operators (human energy source), and the use of fuel in the workers' transportation (commutes to the construction site).

   The building materials, the corresponding total kg of material consumed is determined from the bill of quantities; that amount is then translated into primary energy consumption or embodied energy, and finally expressed in terms of EF.

   The third impact factor is the waste generated on the construction site, which mostly corresponds to the construction and demolition waste (CDW) and, in a smaller amount, to municipal solid waste (MSW).

The last impact factor is the land occupied directly, on which the construction takes place, which consumes land and has a corresponding EF. Therefore, each of the impact factors uses resources (energy, water, manpower, materials) or generates waste.

2. Definition of intermediate elements. Through these elements, consumption is transformed into elements that allow the calculation of various footprints that make up the total EF of the system under study. The intermediate elements are: fuel, electricity, mobility, manufacture, transport and installation of construction materials, CO_2 emissions, and land necessary for the absorption of CO_2 emissions (CO_2 land).

3. Definition of the coefficients. The coefficients in order to transform the resources and its intermediate elements into different footprints (energy, forest, sea, crops, pasture, sea and directly occupied land) are: mobility, transport, embodied energy and waste generation coefficients; transformation factors, such as food performance, waste conversion, emission, absorption, and equivalence factors; and finally, other characteristics include electric mix and forest productivity.

4. Definition of partial and total footprints is given by means of the intermediate elements and the corresponding coefficients, while the partial and global footprints that are generated are obtained in the form of forest, food, energy, built land, and total EF. For these calculations, up-to-date equivalence factors (e), in Table **3** in Chapter 1, must be considered in order to determine the global hectares (gha).

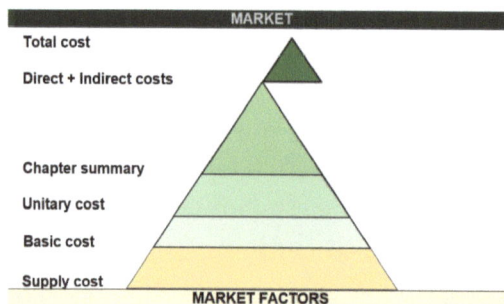

Fig. (24). Pyramidal cost structure [7, 8].

In the methodology, the project's bill of quantities must be used in accordance with a systematic classification breakdown system and its corresponding cost system. For this analysis, the Andalusia Construction Cost Database (ACCD) [6] is used which is a common method to manage information in the region [7, 8]. The breakdown system is hierarchical; each project activity is divided into subgroups of similar characteristics, (see Figs. **24** and **25**).

The total production cost (TPC) covers all production costs incurred by the tasks which are necessary for the projected work, (see Fig. **25**). The basic costs (BC): refers to elements that are a resource: manpower, materials, and machinery; and BC form other more complex costs, such as auxiliary (AC), simple (SUC), complex (CUC), and functional costs (FUC), (see Fig. **25**).

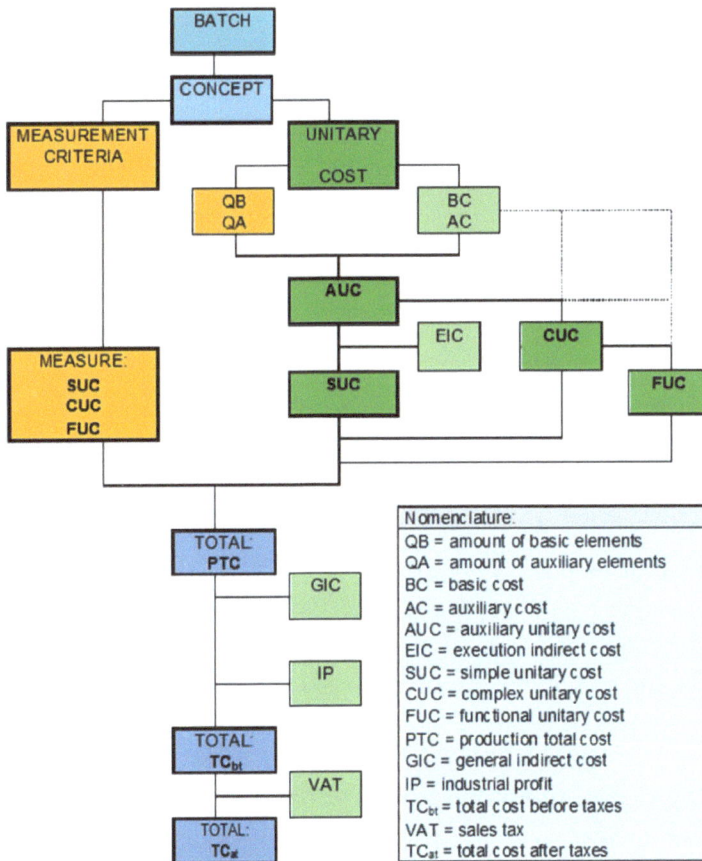

Fig. (25). Budget model [8].

Energy Consumption

Two main types of energy sources are employed on the construction site: fuel and electric power. In order to obtain the fuel consumption in units of volume (litres), first, the total machinery rental costs are determined. The cost of the machinery employed is summarized in Table **36**. First, 20% of those costs is considered fuel consumption. Second, the fuel cost is 1.14 €/litre in the project year, 2008. Finally, the footprint of fuel consumption can be expressed as [4]:

$$EF_{wf} = \frac{F}{EP} \cdot e_f \qquad \qquad \dots \qquad \qquad (13)$$

where:

EF_{wf}: Weighted EF of fuel consumption (gha/year)

F: Fuel consumption (GJ)

EP: Energy productivity of fuel (GJ/ha/year), which is considered as oil (Table **12**). Energy productivity of fuel is expressed as:

$$EP = \frac{A}{E} \qquad \qquad \dots \qquad \qquad (6)$$

where:

A: is the absorption factor and E is the emission factor. A= 5.21 tonnes of CO_2/ha per year [9]

e_f: equivalence factor of forests (gha/ha)

The theoretical consumption of electricity is based on the indirect costs of the project, as explained in Chapter 3. In Table **37** the indirect costs are evaluated for the construction of 107 dwelling project.

Once the electricity consumption in obtained from the combination of empirical data and the indirect cost analysis, as described in Chapter 3; then the electric mix in Andalusia, Spain is used and the corresponding energy productivity (EP) of each source is applied. Renewable sources are assumed to have an irrelevant energy footprint because they have a high EP [10] as compared to fossil fuels and nuclear energy.

Table 36. The rental cost of machinery in the project.

Machinery	Cost (€)
Loader	6,499.56
Dump truck	33,240.06
Backhoe	1,431.73
Bulldozer	22.42
Vibratory roller	4,143.84
Manual mechanical tamper	936.38

Table 37. Determination of the electricity consumption on the construction site.

Code	Concept	Unit	Power consumption	Total Power
Machinery, equipment and tools (per month)			kWh/month	kWh
C122311	Crane	12	1,525	18,300
C122314	Lifting platform	12	305	3,660
C122315	Elevator	12	305	3,660
C12232	Concrete mixer	12	149.45	1,793
C12233	Machine shop of steel reinforcement	12	162.67	1,952
Ancillary and complementary facilities			Power consumption	
C1231	Worksite	m^2	kWh/m^2/year	
C12311	Offices	100	208	20,800
C12312	Meeting rooms	100	208	20,800
C12313	Storage rooms	100	208	20,800
Other electrical consumption		m^2	kWh/m^2/year	
C12531	Lighting of the construction site (plot)	7123.78	1.49	10,614
C12532	Testing facilities (a/c, heating, *etc.*)	7123.78	1.11	7,907
			Total	110,287

The EP for electric power generation is calculated from the EP of fossil and nuclear energy (Table **12**). The efficiency factor is 0.3 [11].

The equation is [4]:

$$EF_{we} = \sum_i \frac{P_i}{EP_i} \cdot e_f \qquad \ldots \qquad \textbf{(11)}$$

where:

EF_{we}: Weighted EF of electricity consumption (gha/year)

P_i: Primary energy consumption (GJ)

EP_i: Energy productivity (GJ/ha/year)

The consumption is calculated from the primary energy consumption *i* and the energy productivity, applied to each of the sources, i.

Water Consumption

The footprint of water consumption is included in the present EF methodology.

The procedure is [4]:

1. Define the water consumption of the work to be analysed which is based on empirical data form 90 dwelling projects which were previously analysed in Chapter 3.
2. Determine the EF. The forest is considered as a water producer, whereby the consumption of this resource is included in that of forest land footprint. In order to calculate forest productivity (m³/ha/year), the hypothesis that a forest of wetlands can produce 1.500 m³ of fresh water per hectare per year is assumed [9].

The equation for the calculation of the EF of water is [4]:

$$EF_{ww} = \frac{W}{FP} \cdot e_f \qquad \ldots \qquad \textbf{(17)}$$

where:

EF_{ww}: Weighted EF of water consumption (gha/year)

W: Water consumption (m³)

FP: Forest productivity (m^3/ha/year)

Food Consumption

In this section, the workers´ food intake is attributed to the EF of the building construction since this activity takes place on the worksite, and is considered as the workers´ "fuel", in the same way as in the methodology developed by Domenech Quesada [9] where business meals are allocated to the Corporate EF. To this end, the total number of manpower hours worked in the project are calculated, which is obtained from the project bill of quantities. The footprint is calculated using the expression [4]:

$$EF_{wfd} = \frac{EF_m}{h_m} \cdot N_h \qquad \ldots \qquad (32)$$

where:

EF_{wfd}: Weighted EF of food (gha/year)

EF_m: EF expressed as gha/year/meal

h_m: 8 hours/meal. One meal per working day is assumed.

N_h: Total number of hours worked

Therefore, it is necessary to obtain the EF_m of the various types of food that make up the daily meal of every worker.

In accordance with the methodology used, four types of footprints are generated:

1. Fossil: generated by all types of food, due to their required processing, or, as in the case of fish, this factor represents the fuel consumed for the capture of the fish. This translates into CO_2 land. The formula is [4]:

$$EF_{mf} = \frac{C \cdot EI}{EP} \cdot e_f \qquad \ldots \qquad (33)$$

where:

EF_{mf}: Fossil EF (gha/year/meal)

C: consumption (t/meal)

EI: energy intensity (GJ/t)

EP: energy productivity (oil) (GJ/ha/year)

2. Pasture: generated by meat and dairy products.
3. Cropland: generated by cereals, beverages, vegetables, sweets, oil and coffee.
4. Productive sea: generated by fish.

For these latter three types of footprint, natural productivities (NP) in Table **17** are used whereby the three types of consumed land are determined: pasture, cropland, and productive sea.

For example, in the case of cropland [4]:

$$EF_{mp} = \frac{C}{NP} \cdot e_p \qquad \qquad \ldots \qquad \qquad (34)$$

where:

EF_{mc}: Cropland EF (gha/year/meal)

NP: Natural productivity (t/ha/year)

e_c: equivalence factor of cropland

Mobility

In order to determine the EF related to the mobility of workers, the following assum-ptions are made [4]:

1. Private vehicles are established as the only means of transport, since it is assumed that the construction work is placed out of the city centre.
2. The average distance travelled by vehicles is 15-30 km.
3. The average vehicle occupancy is 4 people per vehicle. In order to determine the number of workers, the total number of hours worked (calculated in the previous section on food) and the effective duration of the work (in hours) must be known.
4. The average fuel consumption of cars in Spain [11] is used, 7.40 l/100 km.
5. The mobility footprint is determined as in the energy section.

Construction Materials

The footprint of construction materials is determined using the following expression [4]:

$$EF_{wm} = \frac{\sum_i Cm_i \cdot Ese_{mi}}{EP} \cdot e_f \qquad \ldots \qquad (35)$$

where:

EF_{wm}: Weighted EF of construction materials (gha/year)

Cm_i: Material consumption (kg)

Ese_{mi}: Specific embodied energy of material i (MJ/kg)

EP: Energy productivity (oil) (MJ/ha/year)

The footprint of the construction materials is allocated to CO_2-absorption land (forest land). In the case of wooden construction materials, whose quantities are relatively small in the case study, the additional forest footprint is disregarded.

The embodied energy values were obtained by taking the average of various sources [12 - 16], on the condition that no great disparity existed between those values. This embodied energy includes the manufacture, transportation and installation of construction materials, as per the sources consulted.

Based on these values, the consumption of materials (in kilograms) is determined from the bill of quantities of the project studied. Basic Costs (BC) of the ACCD [6] are used. In order to convert units of measurement of BC (m, m^2, m^3, *etc.*) into kg, the coefficients calculated by Mercader [1] are employed (Table **38**).

Table 38. Embodied energy calculation of representative materials in the case study.

	u	M_{mi} (u)	C_{ci}(kg/u)	Cm_i (kg)	Ese_{mi}(MJ/kg)	Ee_{mi}(MJ)
Steel	kg	234,915	1	223,728.87	40.0	8,949,154
Concrete	m^3	1,271	2,500	3,085,849.51	1.0	3,085,849

(Table 38) contd.....

	u	M_{mi} (u)	C_{ci}(kg/u)	Cm_i (kg)	Ese_{mi}(MJ/kg)	Ee_{mi}(MJ)
Bricks	mu	239	1,550	350,373.11	2.9	1,016,082
Gypsum Board	m²	21,253	10	202,413.81	7.0	1,416,896
Cement	t	173	1,000	164,827.65	7.0	1,153,793
Aluminium door	m²	327	20	6,552.00	200.0	1,310,400

The example shown in Table **38** corresponds to specific construction materials which are representative of the case study because of its high volumes consumed. The second column shows the unit in which the material is commercially measured. The remaining columns represent [4]:

M_{mi}: Quantity of material i in the project concerned

C_{ci}: Conversion coefficient of the unit measure of the Basic Cost into weight (kg).

C_{mi}: Consumption of material i (kg)

$$Cm_i = M_{mbi} \cdot C_{ci} \qquad \qquad \dots \qquad \qquad \textbf{(36)}$$

Ese_{mi}: Specific embodied energy of the material i.

Ee_{mi}: Embodied energy of the material i (MJ)

$$Eem_i = Cm_i \cdot Esem_i \qquad \qquad \dots \qquad \qquad \textbf{(23)}$$

In Appendix D, all the construction materials that are part of the project under study are quantified and their corresponding C_{Ci} coefficients (kg/u) are determined. In Table **39** all construction materials which are part of the building construction are grouped together by raw materials.

Table 39. the building directly.

	Weight (kg)	Energy (MJ)
Steel	262,787.11	12,365,307.12
Sand	2,606,574.31	223,470.08
Concrete	14,128,022.79	14,551,548.72
Ceramic	1,484,329.66	4,814,196.15
Plaster	744,397.59	3,910,675.68

(Table 39) contd.....

Cupper	6,229.03	626,682.02
Lime	31,401.52	125,606.10
Cement	171,599.08	1,201,193.53
Porcelain	15,498.95	426,221.13
PVC	12,674.62	1,014,828.01
Wood	53,235.14	160,881.36
Brass	15,355.65	1,535,564.68
Stone	1,268,007.94	2,523,197.48
Aluminium	12,384.97	2,466,611.17
Polyethylene	2,175.72	184,936.23
Polystyrene	1,341.76	147,593.93
Methacrylate	66.75	6,007.50
Paint	147,475.28	10,496,911.66
Glass	10,922.83	196,610.85
Others	-	57,543,346.63
Total	**21,219,156.79**	**114.521.390,01**

Waste

In this section, the environmental impact of waste is evaluated, which is grouped into two main groups: municipal solid waste (MSW), and construction and demolition waste (CDW). All the waste included in the calculation is generated on the construction site by the manpower and the construction materials.

Municipal solid waste is broken down into four types: organic matter, paper/cardboard, plastics, and glass. In the case of the CDW, two types of waste are considered in accordance with the management models in the treatment plants in Andalusia: earth, and mixed CDW. Mixed CDW groups the remains of materials generated during the execution of the work unit and the packaging used in the transport of the materials. In new construction work, excavated earth may represent over 80% of CDW, while the mixed CDW is distributed among the remains of materials and packaging [17].

The determination of the EF of waste is based on the methodology of Wackernagel [18], which states that the footprint associated with waste disposal and emissions is calculated in the same way as for new materials: with the same energy intensity (embodied energy) but subtracting the percentage of energy that can be recovered from recycling. The waste generates fossil footprint.

Conversion rates [9, 19] are used that refer to various types of waste from different origins. For non-hazardous waste, the procedure is based on the energy intensity (EI) of the production of the material from which the waste is made, with a deduction of the percentage of energy that can be recovered by recycling. Non-hazardous waste includes organic, excavated earth, or mixed CDW. The conversion rate is defined as [4]:

$$CR_x = \frac{EI_x}{EP} \cdot (1 - \frac{\%R_x}{100} \cdot \frac{\%SE_x}{100}) \cdot e_f \ ...$$ (26)

where each of these terms is:

CR_x: Conversion rate of non-hazardous waste x (gha/year/t)

EI_x: Energy intensity of the production of the material from which the waste is made, (Table **39**).

EP: Energy productivity of the waste (assumed to be equal to that of fossil fuels).

$\%R_x$: Recycling rate of waste x. In the case of organic waste, nationwide information is used [20], (Table **40**). Data from the Regional Government in Andalusia [21] are used for the recycling rates of paper, plastics, and glass. Of the excavated earth, 50% is reused on site and the remaining 50% is sent to a treatment plant where 80% is recycled. The mixed CDW is also sent to the plant, and a recycling rate of 15% is considered [22].

Table 40. Parameters for the calculation of conversion rates.

	Organic	Paper	Plastics	Glass	Earth	Mixed CDW
EI_x (GJ/t)	20	30	43.75	20	0.10	5
EP (GJ/ha/year)	71	71	71	71	71	71
$\%R_x$	13	50	40	40	80	15

(Table 40) contd.....

	Organic	Paper	Plastics	Glass	Earth	Mixed CDW
%SE$_x$	100	50	70	40	90	90
e$_f$ (gha/ha)	1.34	1.34	1.34	1.34	1.34	1.34
NP (t/ha/year)		1.01				

%SE$_x$: percentage of energy recovered by recycling.

Finally, the total waste footprint is [4],

$$EF_{wws} = \sum_i CR_{x_i} \cdot G_i \qquad \ldots \qquad (29)$$

where:

EF$_{wws}$: Weighted EF of the waste (gha/year)

CR$_{xi}$: Weighted Conversion Rate (gha/year/t)

G$_i$: Waste Generation (t)

The procedure is as follows [4]:

1. Determination of MSW and CDW quantities. These calculations are based on statistical data [21, 23], and on a software tool [17, 24], respectively.
2. Determination of conversion rates (CR) for each waste type.
3. Calculation of the waste EF (forestry and energy footprint).

Determination of the Built Land EF

The built land footprint is obtained from the land area used by the urbanization and by the building under study, whose allocated surface is in the form of "productive land used directly", as previously defined in the proposed EF methodology.

The EF calculation is [4],

$$EF_{wb} = S.e_b \qquad \ldots \qquad (31)$$

where:

EF_{wb}: Weighted EF of built land (gha/year)

S: Surface area consumed (ha/year)

e_b: Equivalence factor of built land

The land used directly is considered to have the same productivity as that of agricultural land, since most of the infrastructure and built environment are located on agricultural productive land.

RESULTS

Energy Consumption EF

The fuel consumption is determined from the rental costs of the machinery listed in the bill of quantities where 20% are considered fuel cost. Heavy machines and small combustion engine machines are used during the construction, urbanization and the indirect activities included as indirect costs. The results are summarized in Table **41**.

Table 41. Costs, consumption and EF of machinery.

Machinery	Rental cost (€)	Fuel cost (€)	Consumption (l)	EF of fuel (gha/year)
Building	167,708.63			
Urbanization	16,588.42			
Indirect Costs	71,152.76			
Total cost	255,449.81	51,089.96	44,760.79	29.57

The electricity consumption in kWh is obtained from the indirect cost analysis, (Table **36**). In order to obtain the electricity footprint, it is also necessary to determine the sources of electric power in Spain [25]. Only the footprint from fossil fuels and nuclear energy is considered, as previously explained. The results are shown in Table **42**.

Table 42. Electricity EF.

Electricity consumption (kWh)	110,287
Electricity consumption (GJ)	397
Efficiency factor	0.30
Primary energy consumption (GJ)	1,323
Electricity EF(fossil) (gha/year)	19.83

Water Consumption EF

The water consumption, obtained from other similar dwelling construction projects that have already been executed, is 1,536.55 m³, thereby resulting in an EF (forest footprint) of 1.38 gha/year.

Food Consumption EF

In order to calculate the food consumption EF, the total number of manpower hours worked during the entire project is calculated which obtained from the project bill of quantities. In Tables **43**, **44** and **45** a detail description of the workers per speciality and/or activity is presented. The corresponding percentages are represented in Fig. (**26**).

Table 43. Manpower hours during the urbanization works.

Labourer	Working hours
Gardener	3,924.31
Regular labourer	680.23
1st class bricklayer	25,141.92
Specialist labourer	1,822.21
1st class labourer	10,497.40
Plumber	588.760
Painter	4,372.77
Welder	5,279.76
Electrician	8,580.00
Total manpower hours	**4,280.57**

Table 44. Manpower hours during the building construction.

Labourer	Working hours
Carpenter	3,924.31
Locksmith	680.23
Specialist labourer	25,141.92
Regular labourer	1,822.21
1st class specialist	10,497.40
2dn class specialist	588.76
Reinforced steel labourer	4,372.77
Concrete casting labourer	5,279.76
1st class bricklayer	8,580.00
2st class bricklayer	58.79
Specialist assistant	491.04
Installer	1,526.09
Assistant	708.46
Heating installer	1,825.65
Plumber	4,400.47
Electrician	3,330.41
Installer assistant	822.50
Furniture installer	2,931.11
Water proofing installer	353.85
Welder	991.00
Painter	4,903.22
Glazier	490.76
Tyler	2,268.48
Welder assistant	5,058.11
False ceiling placement	3,008.89
Plasterer	4,629.87
Total manpower hours	**98,686.05**

Table 45. Manpower hours which are part of the project indirect costs.

Labourer	Unit	Quantity	Unit cost (€/unit)	Cost (€)
Manager	month	12.00	2,805.53	33,666.36

(Table 45) contd.....

Labourer	Unit	Quantity	Unit cost (€/unit)	Cost (€)
Foreman	month	12.00	2,437.50	29,250.00
Storeroom keeper	month	6.00	2,437.50	14,625.00
Security personnel	month	6.00	2,681.28	16,087.68
Presence control	month	6.00	2,681.28	16,087.68
Internal waste transport	m^2	16,000.00	1.66	26,560.00
Internal cleaning and waste picking	m^2	16,000.00	2.56	40,960.00
Tool and material transport	m^2	16,000.00	0.69	11,040.00
Crane installation and uninstallation	unit	2.00	5,703.63	11,407.26
Technical support	month	12.00	3,212.05	38,544.60
Administration	month	12.00	2,087.60	25,051.20
Power and water connections and lines	unit	1.00	2,188.31	875.34
Provisional roads	unit	1.00	799.57	319.83
Total cost (€)				264,474.95
Average labourer cost (€/h)				16.70
Manpower (h)				15,836.82

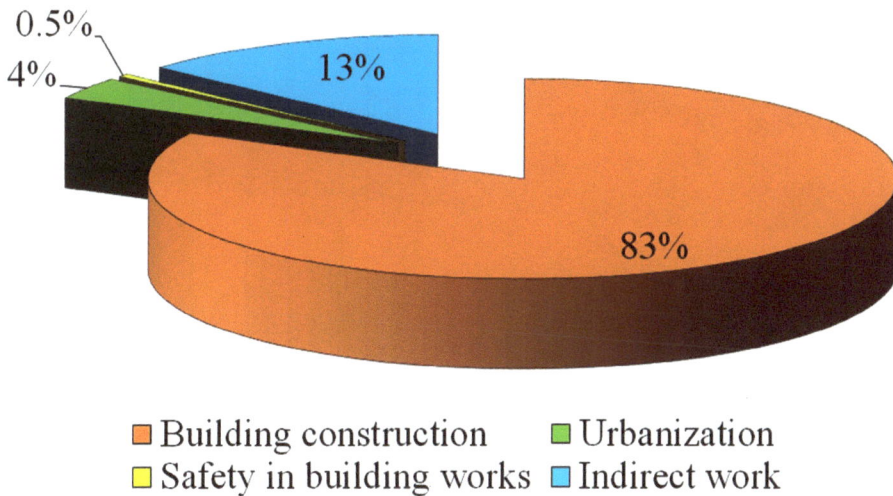

Fig. (26). Manpower hour distribution (%).

The results are summarized in Table **46**, and include the manpower that is part of the indirect costs and workplace health and safety (WHS).

Table 46. Total cost of manpower.

Task	Manpower Hours
Building	98,686.05
Urbanization	4,280.57
Building WHS	604.46
Urbanization WHS	10.93
Indirect Costs	15,836.82
Total	119,418.84

The EF_m from the different types of food that make up the daily meals of the workers are then obtained, by using the data in Table **20**. The results are shown in Fig. (**27**).

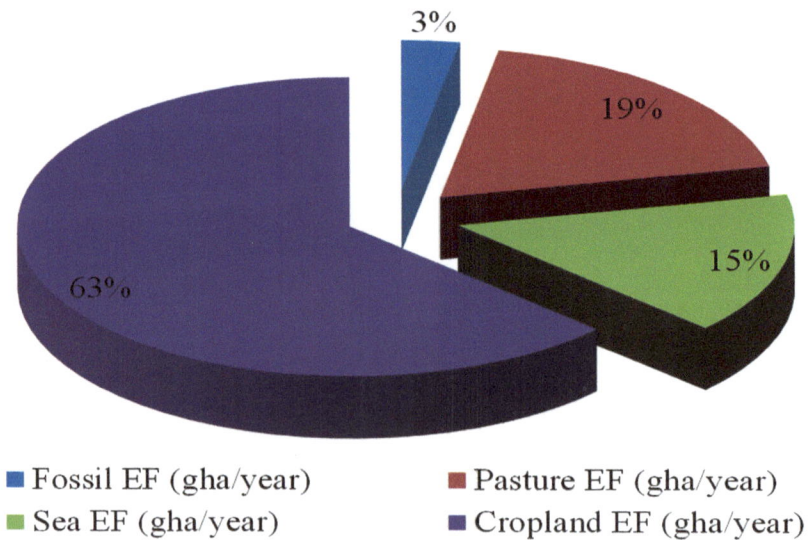

3%
19%
15%
63%

- Fossil EF (gha/year)
- Pasture EF (gha/year)
- Sea EF (gha/year)
- Cropland EF (gha/year)

Fig. (27). EF of food.

Mobility EF

The EF of mobility is obtained from the total number of working hours Table **46** and working days per year; that results in the number of workers each day in the construction site. Also the average car occupancy and distance travelled are needed; and a total distance travelled in a year is obtained (Table **47**). Finally, the total distance is multiplied by the average fuel consumption and the EF of mobility (fossil) is 6.194 gha/year (Table **47**).

Table 47. The EF of mobility.

Total working hours (h)	119,418.84
Total construction work duration (h)	1533
Number of workers Nt	77.9
Real number of workers	78
Vehicle occupancy	4
Number of vehicles	19,5
Real number of vehicles	20
Average distance travelled (km)	30
Total distance travelled each day (km)	600
Working days during one year	**192**
Total distance travelled (km)	**115200**
Average consumption (l/100 km)	7.4
Total fuel consumption (l)	8,524.8
Maintenance (10%)	1.1
Conversion factor (MJ/l)	35
Energy productivity (MJ/ha)	71,000
EF of fuel consumption (ha)	4.623
Equivalence factor (hag/ha)	1.34
EF of mobility (hag)	**6.194**

Construction Materials EF

The total embodied energy of construction materials in the building and its urbanization are, respectively, 114,521,390.01 MJ and 10,313,394.16 MJ. Their total is 124,834, 784.18 MJ. In Fig. (**28**), the embodied energy of the main construction materials in the project is represented. Five construction materials, concrete, steel, paint, ceramics, and gypsum, represent 40% of the total. The EF of construction materials is 2,357.77 gha/year (fossil footprint).

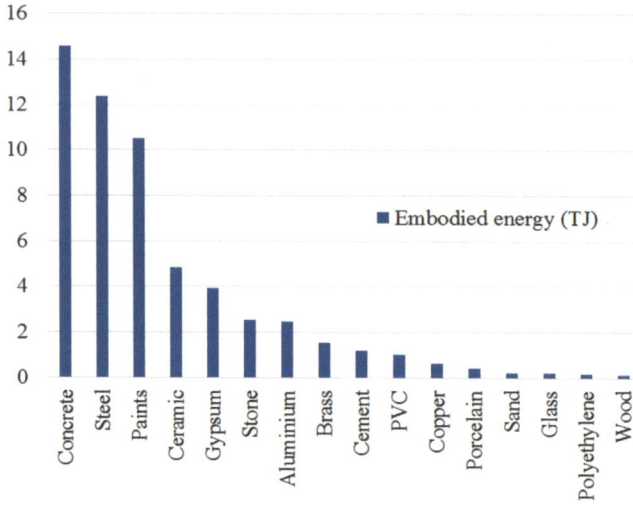

Fig. (28). Embodied energy in the most representative construction material in the project.

Waste EF

The generation of MSW and CDW are determined through statistical databases and software tools, respectively. In the case of CDW, the software tool employed determines 22,400 m³ of excavated earth (of which 50% is reused) and 1,920 m³ of mixed CDW. The results are shown in Table **48**. In Fig. (**29**), the percentages of each waste type that contributes towards the total EF are represented.

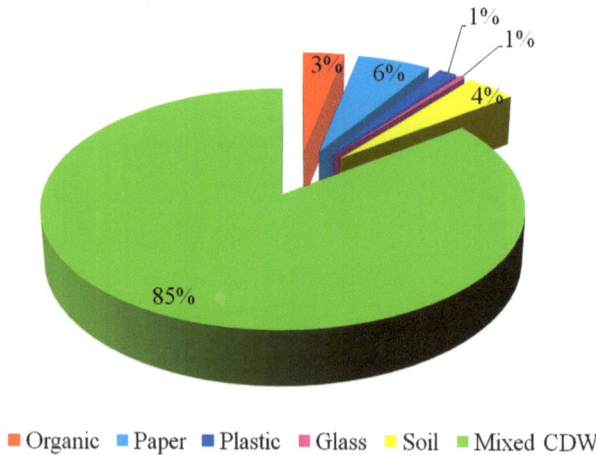

Fig. (29). Waste EF (%).

Table 48. Waste EF.

	Organic	Paper	Plastic	Glass	Earth	Mixed CDW	Total EF
G (t)	17.71	8.45	4.43	2.83	13.440	1.920	
CR (gha/year/t)	0.3284	1.2207	0.5945	0.3171	0.0005	0.0816	
EF (fossil) (gha/year)	5.82	10.32	2.63	0.89	7.10	156.72	183.48

Built Land EF

In order to determine the built land EF, the total land area must be identified. To this end, the surface area for blocks and surface area for roads are computed, giving a total area of 7,123.78 m^2, (see Table **49**). By means of an equivalence factor of 2.21 gha/ ha (agricultural land), the EF of built land can be calculated as 1.57 gha/year.

Table 49. Land occupy directly by roads and buildings.

Land area	m^2
Lot 1	1,484.80
Lot 2	1,320.08
Lot 3	4,318.90
Total	**7,123.78**

Total EF

Tables **50 and 51** show the overall results, expressed in gha/year and gha/year/m^2. Fig. (**30**) represents the footprint distribution per source of impact. In Table **51**, the constructed area considered is that of blocks, not the built land (which includes roads). Therefore, the total area is obtained adding the parking, ground, first, second and third floor areas, 15,820.03 m^2 (Table **34**).

Moreover, a sensitive analysis should be performed in order to observe the behaviour of the variables. For example, two models of CDW management are compared. First, neither is the CDW sorted for recycling, nor is the soil reused, and therefore the footprint is 2,004.82 gha. In a second scenario, the waste is sorted and recycled and the soil reused, as in the original case study, and the resulting EF is 183.48 gha. The indicator is sensitive to changes in its variables.

Table 50. EF of 107 dwellings construction.

	Footprint type (gha/year)					
Impact	**CO_2-absorption land**	**Forest**	**Pasture**	**Productive sea**	**Cropland**	**Built land**
Machinery	29.57					
Electricity	19.83					
Water		1.37				
Food	2.58		17.22	14.10	58.05	
Mobility	6.19					
Materials	2,357.77					
Waste	183.48					
Built land						1.57
Totals	2,591.53	1.37	17.22	14.10	58.05	1.57
GRAND TOTAL	**2,684.60**					

Table 51. Total EF (with respect to constructed area).

	Footprint type (gha/year/m^2)					
Impact	**CO_2-absorption land**	**Forest**	**Pasture**	**Productive sea**	**Cropland**	**Built land**
Machinery	0.001869					
Electricity	0.001254					
Water		0,000134				
Food	0.000163		0,001089	0,000891	0,003669	
Mobility	0.000392					
Materials	0.148927					
Waste	0.011598					
Built land						0,000100
Totals	0.163814	0,000134	0,001089	0,000891	0,003669	0,000100
GRAND TOTAL	**0.17**					

Other possible changes in the scenarios can take place with the construction materials. For the same constructive solutions, such as reinforced concrete, recycled steel and concrete can substitute the original materials in the project. The

EF of the resulting construction materials is changed from the original 2,356.04 gha to an improved scenario of 1,965.40 gha.

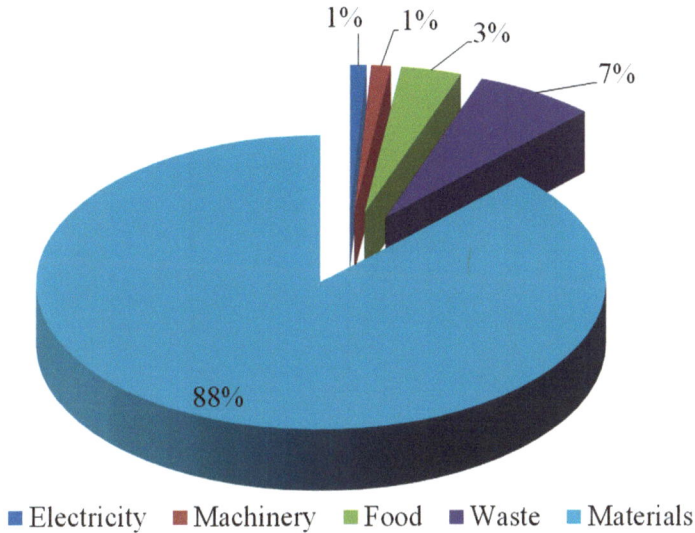

Fig. (30). Total EF (%).

EF of Ten Different Dwelling Projects

In order to determine if the EF methodology presented is sensitive to changes in project characteristics, ten dwelling projects of different size and configuration, are assessed. Those projects were analysed by Gonzalez Vallejo *et al*. [27] with the original model of Solis-Guzman *et al*. [4]. The present results are compared to those previous analyses.

The projects main features are listed in Tables **52** and **53** where the total number of rooms includes bedrooms and kitchen but excludes bathrooms. Other characteristics which are common to the ten projects are:

1. The windows have aluminium frames.
2. The interior doors and shades are made of wood.
3. All dwellings have tap water, city sewerage system and domestic hot water.
4. Electricity is the main energy source.

Table 52. Dwelling types (projects 1 to 5) [27].

DESCRIPTION		PROJECT				
		1	2	3	4	5
Dwelling type		Detached	Semidetached	Two or more dwellings per building		
Surface per dwelling		181	143	72		
Over ground floors		1	2	2	3	3
Underground floors		0	0	1	1	1
Construction	Vertical structure	Load-bearing walls	Reinforced concrete			
	Horizontal structure	Concrete slab	Concrete slab	One-direction forging		
	Roof	Pitched	Flat	Flat	Pitched	Flat
	Exterior wall	Continuous plaster	Stones	Others	Ceramic brick	Ceramic brick
Interior finishes	Floor tile	Ceramic		Stone	Wood	Ceramic
	False ceiling	No		Yes		
Rooms		6		4		
Bathrooms		3	2			
Installations	Heating	Yes			No	Yes
	A/C	Yes	No			
	Elevator	No			Yes	
	In-situ water treatment	No				
Energy	Natural gas	No	Yes	No	Yes	No
	Thermal solar	Yes	No	Yes	No	Yes
Garage		No		Yes		

Table 53. Dwelling types (projects 6 to 10) [27].

DESCRIPTION	PROJECT				
	6	7	8	9	10
Dwelling type	Two or more dwellings per building				
Surface per dwelling	70	72			
Over ground floors	4	5	5	6	10

(Table 53) contd.....

DESCRIPTION		PROJECT				
		6	7	8	9	10
Underground floors		2	1	2	1	2
Construction	Vertical structure	Reinforced concrete				
	Horizontal structure	Concrete slab	One-direction forging			
	Roof	Flat	Pitched	Flat	Pitched	Flat
	Exterior wall	Continuous plaster	Stones	Others	Ceramic brick	Ceramic brick
Interior finishes	Floor tile	Stone	Ceramic	Wood	Ceramic	Wood
	False ceiling	Yes				
Rooms		6	4			
Bathrooms		2				
Installations	Heating	Yes	No		Yes	No
	A/C	No			Yes	No
	Elevator	Yes				
	In-situ water treatment	No				Yes
Energy	Natural gas	No		Yes	No	Yes
	Thermal solar	Yes		No	Yes	No
Garage		Yes				

Table 54. Resources consumed and waste generated in projects 1 to 5 [27].

Resources per square meter constructed		Project number				
		1	2	3	4	5
Weight (kg)	Steel	24.32	18.77	21.40	20.66	20.66
	Concrete	2120.10	1281.23	1219.22	1262.09	1262.09
	Ceramic	444.99	518.58	405.35	293.38	344.42
	Others	468.00	304.74	297.48	295.08	289.72
	Total	3057.40	2123.32	1943.46	1871.21	1916.90
Embodied energy (MJ)	Steel	972.67	750.83	855.81	826.40	826.40
	Concrete	2132.52	1288.68	1225.54	1266.60	1266.60
	Ceramic	1328.71	1535.65	1198.62	876.79	1024.81
	Others	4018.06	3087.31	2463.68	2213.95	2223.22
	Total	8451.96	6662.47	5743.66	5183.75	5341.04
Manpower	Working (h)	13.02	11.62	9.89	8.99	9.35
Machinery	Rental (€)	2.96	1.95	6.49	4.95	4.94

(Table 54) contd.....

Resources per square meter constructed		Project number				
		1	2	3	4	5
Utilities	Power (€)	19.46	19.46	19.46	19.46	19.46
	Water (m³)	0.15	0.11	0.12	0.06	0.06
Waste (t)	Urban	5.10E-03	4.48E-03	3.78E-03	3.48E-03	3.61E-03
	Construction	0.19	0.17	0.13	0.15	0.15
Occupied land (ha)		1.00E-04	5.00E-05	5.00E-05	3.33E-05	3.33E-05

The resources consumed per square meter constructed in each project are listed in Tables **54** and **55**. Concrete, steel and ceramic materials are the most consumed in all cases. The main differences from Gonzalez-Vallejo *et al.* [27] previous work are the electricity consumption calculation and the manpower food consumption. In the present analysis the electricity is obtained from empirical data; the average power consumption per square meter constructed is defined from 30 already built projects. The food consumption has been calculated from new data obtained from FAO and EFN as established in Chapter 4. The EF of food and electricity are significantly reduced in the present model (about 90% smaller) with respect to previous results [4, 27, 28].

Table 55. Resources consumed and waste generated in projects 6 to 10 [27].

Resources per square meter constructed		Project number				
		6	7	8	9	10
Weight (kg)	Steel	51.88	16.34	45.27	14.64	40.31
	Concrete	1066.17	1055.19	1007.99	1069.33	964.28
	Ceramic	297.14	356.57	300.44	369.74	310.84
	Others	284.45	265.82	289.29	251.22	288.69
	Total	1699.65	**1693.92**	**1642.99**	**1704.94**	**1604.12**
Embodied energy (MJ)	Steel	2075.36	653.79	1810.79	585.71	1612.41
	Concrete	1070.69	1058.35	1012.50	1071.14	968.79
	Ceramic	886.95	1062.92	896.54	1103.28	926.70
	Others	2073.13	2255.37	2106.07	2374.03	2125.39
	Total	6106.13	**5030.43**	**5825.90**	**5134.15**	**5633.29**
Manpower	Working (h)	9.06	9.14	9.05	0.44	0.41
Machinery	Rental (€)	6.84	3.52	5.89	2.41	5.17

(Table 55) contd.....

Resources per square meter constructed		Project number				
		6	7	8	9	10
Utilities	Power (€)	19.46	19.46	19.46	19.46	19.46
	Water (m³)	0.09	0.03	0.03	0.03	0.03
Waste (t)	Urban	3.44E-03	3.44E-03	3.47E-03	3.52E-03	3.40E-03
	Construction	0.12	0.14	0.12	0.15	0.10
Occupied land (ha)		2.50E-05	2.00E-05	2.00E-05	1.00E-05	1.00E-05

The resources consumed and waste generated have the EF listed in Tables **54** and **55**. The EF per square meter constructed remains constant for multi-family buildings of three storeys or more, that corresponds to projects 3 to 10 in Tables **56** and **57**. The EF of detached dwellings is significantly higher (50%) than all multi-family projects analysed. The first project analysed of 107 dwellings has a higher EF than project no. 6, with a similar characteristics, because the first includes the road paving, gardening and indirect cost analysis, and this data is not available for the other 10 projects.

Table 56. EF calculation of the ten dwelling projects [27].

EF (hag/m²)	Project number				
	1	2	3	4	5
Energy	1.63E-01	1.29E-01	1.13E-01	1.02E-01	1.05E-01
Forrest	1.28E-04	1.28E-04	1.28E-04	1.28E-04	1.28E-04
Pasture	2.46E-03	1.97E-03	1.97E-03	1.47E-03	1.97E-03
Sea	1.77E-03	1.77E-03	1.77E-03	1.18E-03	1.18E-03
Crops	9.15E-03	9.15E-03	4.58E-03	4.58E-03	4.58E-03
Occupied land	2.51E-04	1.26E-04	1.26E-04	8.37E-05	8.37E-05
Total EF (hag/m²)	**0.18**	**0.14**	**0.12**	**0.11**	**0.11**

Table 57. EF calculation of the ten dwelling projects [27].

EF (hag/m²)	Project number				
	6	7	8	9	10
Energy	1.20E-01	9.88E-02	1.14E-01	1.01E-01	1.10E-01
Forrest	1.28E-04	1.28E-04	1.28E-04	1.28E-04	1.28E-04

(Table 57) contd.....

EF (hag/m²)	Project number				
	6	**7**	**8**	**9**	**10**
Pasture	1.47E-03	1.97E-03	1.47E-03	1.97E-03	1.47E-03
Sea	1.18E-03	1.18E-03	1.18E-03	1.18E-03	1.18E-03
Crops	4.58E-03	4.58E-03	4.58E-03	4.58E-03	4.58E-03
Occupied land	6.28E-05	5.02E-05	5.02E-05	2.51E-05	2.51E-05
Total EF (hag/m²)	0.13	0.11	0.12	0.11	0.12

The energy footprint includes the impact produce by the embodied energy of construction materials, the food processing energy, workers transportation, electricity and waste (MSW and CDW). The most important resources are the construction materials which represent 88 to 90 % of the total EF in all projects.

The food intake by the construction workers on the construction site generates other three footprints: pasture, sea and cropland. This EF of food is 5 to 9% of the total.

Another significant footprint is due to waste generate on the construction site which is divided between CDW (70% approx.) and MSW (30% approx). This footprint accounts for the 2-3% of the total.

In all the projects studied, the remaining footprints are not significant, the forest land used to produce water is less than 1% and the directly occupied land is also less than 1% of the total global hectares consumed by the construction.

CONCLUDING REMARKS

Eleven different dwelling projects built in Spain have been assessed with the proposed methodology. The impact factors analysed are: direct consumption (energy and water), indirect consumption (manpower and construction materials), waste generation, and built land. The manpower consumption in building construction produces food expenditure by the operators and the use of fuel derived from worker mobility (trips to the construction site).

For EF of construction materials, a quantitative study is performed on the building materials; that amount is then translated into resources expressible in terms of EF.

The third factor is the impact of waste generated on the construction site. The last impact factor is the land on which the construction itself is built. Therefore, each of the impact factors either uses resources (energy, water, manpower, materials) or generates waste.

The partial and global footprints that are generated in the construction phase of the dwelling sector are obtained (gha/year/m^2): CO_2-absorption land is over 95% in all cases analysed, cropland is 3-2%; and the remaining categories of forest, pasture, productive sea, and built land are only 1-2% of the total. The construction materials represent 88% of the total EF. If all manpower-related EFs are grouped together (mobility, food and MSW), this EF represents 5-9% of the total. And waste (CDW and MSW) represents 2-3%. Strategies that reduce the embodied energy of construction materials can reduce the first of these three EFs significantly, as established in the case study. The second, manpower-related, EF can be reduced by incorporating constructive solutions that are more industrialized, since they are less labour intensive. And the third can be reduced with on-site recycling or reuse of materials.

The difficulties in the analysis of the EF of projects indirect cost has been successfully overcome by following the same methodology as that applied to the direct costs. To this end, all work units or activities that take place on the construction site can be grouped into: manpower, machinery, and/or construction materials; and the corresponding calculation procedure can be applied.

Based on the overall results, on one hand, it is clearly noticeable that the most important footprint is CO_2-absorption land. Within this footprint, the construction materials are highly significant which represent approximately 90% of the total EF in the eleven projects analyzed. On the other hand, the footprint of water usage and land consumption has little appreciable contribution to the total.

The EF indicator has proven to be sensible to changes in the buildings characteristics such as being detached or semi-detached dwellings or the number of storeys per building. Detached houses have the 50% more EF than multi-family buildings. The EF per square meter constructed remains constant in buildings with three or more storeys.

CONFLICT OF INTEREST

The author confirms that this chapter has no conflict of interest.

ACKNOWLEDGEMENT

Ministry of Innovation and Science, through the concession of the R+D+I project: Evaluation of the EF of construction in the residential sector in Spain. (EVAHLED). 2012-2014. *Ministerio de Innovación y Ciencia, por la concesión del Proyecto I+D+i: Evaluación de la huella ecológica de la edificación en el sector residencial en España (EVAHLED). 2012-2014.*

DISCLOSURE

"Part of this chapter has been previously published in Ecological Indicators. Volume 25, February 2013, Pages 239–249 doi:10.1016/j.ecolind.2012.10.008".

REFERENCES

[1]   P. Mercader, *Cuantificación de los recursos consumidos y emisiones de CO_2 producidas en las construcciones de Andalucía y sus implicaciones en el Protocolo de Kyoto. (Quantification of the resources consumed and of CO_2 emissions on the construction sites of Andalusia and its implications for the Kyoto Protocol).* Ph. D. thesis, Universidad de Sevilla, Seville, Spain, 2010. http://fondosdigitales.us.es/tesis/tesis/1256/cuantificacion-de-los-recursos-consumidos-y-emisiones-de - co2-producidas-en-las-construcciones-de-andalucia-y-sus-implicaciones- en-el-protocolo-de-kioto/.

[2]   P. Mercader, M. Marrero, J. Solís-Guzmán, M.V. Montes, and A. Ramírez de Arellano, "Cuantificación de los recursos materiales consumidos en la ejecución de la Cimentación (Quantification of material resources consumed during concrete slab construction)", *Informes de la Construcción,* vol. 62, pp. 125-132, 2010. [http://dx.doi.org/10.3989/ic.09.000]

[3]   E. Holden, "Ecological Footprints and Sustainable Urban Form", *J. Housing Built Environ.,* vol. 19, pp. 91-109, 2004. [http://dx.doi.org/10.1023/B:JOHO.0000017708.98013.cb]

[4]   J. Solís-Guzmán, M. Marrero, and A. Ramírez-de-Arellano, "Methodology for determining the ecological footprint of the construction of residential buildings in Andalusia (Spain)", *Ecol. Indic.,* vol. 25, pp. 239-249, 2013. [http://dx.doi.org/10.1016/j.ecolind.2012.10.008]

[5]   J. Solís-Guzmán, A. Martínez-Rocamora, and M. Marrero, "Methodology for determining the carbon footprint of the construction of residential buildings", In: S.S. Muthu, Ed., *Assessment of Carbon Footprint in Different Industrial Sectors, Volume 1.* Springer Science + Business Media: Singapore,

2014, pp. 49-83.
[http://dx.doi.org/10.1007/978-981-4560-41-2_3]

[6]   Andalusia Construction Cost Database (ACCD), *Base de Costes de la Construcción de Andalucía, 2008. Consejería de Obras Pública y Vivienda de la Junta de Andalucía.* 2008. [Online] Available: http://www.juntadeandalucia.es/obraspublicasyvivienda/portalweb/web/areas/vivienda/texto/bcfbb3af-ee3a-11df-b3d3-21796ae5a548. [Accessed Sept 15, 2011].

[7]   M. Marrero, and A. Ramirez-de-Arellano, "The building cost system in Andalusia: application to construction and demolition waste management", *Construct. Manag. Econ.,* vol. 28, pp. 495-507, 2010.
[http://dx.doi.org/10.1080/01446191003735500]

[8]   M. Marrero, A. Fonseca, R. Falcon, and A. Ramirez-de-Arellano, "Schedule and cost control in dwelling construction using control charts", *Open Constr. Build. Technol. J.,* vol. 8, pp. 63-79, 2014.
[http://dx.doi.org/10.2174/1874836801408010063]

[9]   J.L. Domenech Quesada, *Huella ecológica y desarrollo sostenible (Ecological Footprint and Sustainable Development).* AENOR: Madrid, Spain, 2007.

[10]   M. Wackernagel, and W. Rees, *Our Ecological Footprint: Reducing Human Impact on the Earth.* New Society: British Columbia, Gabriola Island., 1996.

[11]   Instituto para la Diversificación y Ahorro de la Energía (IDAE), *Guía Práctica de la Energía: Consumo Eficiente y Responsable (Practical Energy Guide: Efficient and Responsible Consumption).* IDAE: Madrid, Spain, 2007. [Online] Available: http://www.idae.es/index.php/mod.documentos/mem.descarga?file=/documentos_11406_Guia_Practica_Energia _3ed_A2010_509f8287.pdf. [Accessed Sept 18, 2014].

[12]   M. Nye, and Y. Rydin, "The contribution of ecological footprinting to planning policy development: using REAP to evaluate housing policies for sustainable construction", *Environ. Plann. B Plann. Des.,* vol. 35, no. 2, pp. 227-247, 2008.
[http://dx.doi.org/10.1068/b3379]

[13]   A. Cuchí, *Arquitectura y sostenibilidad (Architecture and Sustainability).* Universidad Politécnica de Cataluña (UPC): Barcelona, Spain, 2005.

[14]   B. Berge, *The Ecology of Building Materials.* Architectural Press: Amsterdam, Holland, 2009.

[15]   ITEC, *Metabase-TCQ 2000: Datos Ambientales (Environmental Data).* ITEC: Barcelona, Spain, 2005. [Online] Available: http://www.itec.es/programas/tcq/. [Accessed Sept 15, 2011].

[16]   General Directorate of Housing, Architecture and Urbanism, Cerdá Institute, and IDAE, *Guide to Sustainable Building, Energy and Environmental Quality in Construction.* Ministry of Development: Madrid, Spain, 1999.

[17]   J. Solís-Guzmán, M. Marrero, M.V. Montes-Delgado, and A. Ramírez-de-Arellano, "A Spanish model for quantification and management of construction waste", *Waste Manag.,* vol. 29, no. 9, pp. 2542-2548, 2009.
[http://dx.doi.org/10.1016/j.wasman.2009.05.009] [PMID: 19523801]

[18]   M. Wackernagel, R. Dholakia, D. Deumling, and D. Richardson, *Redefining Progress, Assess your Household's Ecological Footprint. v 2.0,* 2000. [Online] Available: http://greatchange.org/ng-

footprint-ef_household_ evaluation.xls. [Accessed Sept 15, 2011].

[19]   E. Marañón, G. Iregui, J.L. Domenech, Y. Fernández Nava, and M. González, "Propuesta de índices de conversión para la obtención de la huella de los residuos y los vertidos. (Proposed conversion rates to obtain the footprint of waste and effluents)", *Observatorio iberoamericano del desarrollo local y la economía social,* vol. 1, no. 4, April-June 2008.

[20]   Observatorio de la Sostenibilidad (Observatory of Sustainability in Spain), "Sustainability in Spain 2007", *Available: www.sostenibilidad -es.org,* 2008. [Online]

[21]   Andalusia Ministry of Environment, *Environment Report 2008.* Andalusia, Spain, 2009. [Online] Available: http://www.juntadeandalucia.es/medioambiente/site/web/menuitem.318ffa00719ddb10e89 d04650525ea0/?vgnex toid=3b32db0dee134210VgnVCM1000001325e50aRCRD. [Accessed Sept 15, 2011].

[22]   Gremio de Entidades del Reciclaje de Derribos (GERD), *IV Congreso Nacional de Demolición y Reciclaje (IV National Congress of Demolition and Recycling)*, May 20-22, 2009., Zaragoza, Spain.

[23]   Spain ME (Ministry of Environment), *Plan Nacional de Residuos de Construcción y Demolición 2001–2006 (National C&D Waste Plan 2001–2006).* Ministry of the Environment: Madrid, Spain, 2001.

[24]   A. Ramirez-de-Arellano Agudo, J. Solís-Guzmán, and J. Pérez Monge, Generación de RCD versión 2.0 (Software de Evaluación de RCD para Tramitación de Licencias Municipales) (CDW Generation 2.0: CDW Evaluation Software for Processing of Municipal Licenses), Universidad de Sevilla, Spain, 2008.

[25]   Spain MI (Ministry of Industry), *Estructura de generación eléctrica en España. La Energía en España 2007 (Structure of generation of electricity in Spain. Energy in Spain 2007),* 2008. [Online] Available: http:// www.aven.es/ pdf/la_energia_en_espana_2007.pdf [Accessed Sept 18, 2014].

[26]   M. Calvo, *Informe de Síntesis: Análisis Preliminar de la Huella Ecológica en España. (Synthesis Report: Preliminary Analysis of the Ecological Footprint in Spain).* Ministerio de Medio Ambiente: Spain, 2007.

[27]   P. Gonzalez-Vallejo, J. Solis-Guzman, and M. Marrero, "La construcción de edificios residenciales en España en el período 2007- 2010 y su impacto según el indicador huella ecológica", (The constructionof residential buildings in Spain in the period 2007-2010 and its impact according to the EcologicalFootprint indicator)", *Informes de la Construcción,* vol. 67, pp. 539-552, . [http://dx.doi.org/10.3989/ic.14.017]

[28]   P. González-Vallejo, M. Marrero, and J. Solís-Guzmán, "The ecological footprint of dwelling construction in Spain", *Ecol. Indic.,* vol. 52, pp. 75-84, 2015. [http://dx.doi.org/10.1016/j.ecolind.2014.11.016]

APPENDIX A. NOMENCLATURE

Variable	Expression	Meaning	Measuring unit
1	AA	Appropriate area for the production of each category	ha
2	C	Total Consumption	t, m^3, GJ
3	P	Productivity	t/ha, GJ/ha
4	aa	Appropriate area for the production of each category per habitant	ha/hab
5	N	Size of analyzed population	hab
6	ef_N	EF	ha/hab y year
7	EF	EF	ha/year
8	EF_W	Weighted EF	gha/year
9	e	Equivalence factor	gha/ha
10	Lsp	Standard production land	gha
11	Lp	Productive land	ha
12	Y	Yield factor	-
13	Lpc	Corrected productive land	gha
14	Ltp	Total productive land	gha
15	Lb	Land for biodiversity	gha
16	D	Ecological deficit	gha
17	Pee	Electric energy production	GJ
18	Fef	Efficiency	-
19	EP	Fuel productivity	GJ/ha
20	A	Absorption factor	$kg\ CO_2/ha$
21	E	Emission factor	$kg\ CO_2/GJ$
22	t	Time Lapse	year
23	FP	Forest productivity	m^3/ha
24	EF_{ww}	Weighted EF of water consumption	gha
25	e_f	Forest equivalency factor	gha/ha
26	NP	Natural productivity	t/ha
27	EF_{wf}	Weighted EF of food (fossil)	gha
28	EI	Energy intensity	GJ/t
29	E	CO_2 emissions	kg
30	Fc	Conversion factor	-

Table contd.....

Variable	Expression	Meaning	Measuring unit
31	Eemi	Embodied energy	MJ
32	Cmi	Material consumption	kg
33	Esemi	Specific energy embodied of material	MJ/kg
34	CR	Conversion rate of non-hazardous waste	ha/t
35	EPi	Energy productivity	GJ/ha
36	%Rx	Recycling rate	-
37	%SEx	Percentage of energy saved by recycling	-
38	EFwws	Weighted EF of the waste	gha
39	CRx	Weighted conversion rate of non-hazardous waste	gha/t
40	e_f	Fossil energy equivalence factor	gha/ha
41	EF_{wb}	Weighted EF of built land	gha
42	S	Surface area consumed	ha
43	e_b	Equivalence factor of built land	gha/ha

APPENDIX B. ABBREVIATIONS

1. LCA: Life Cycle Analysis
2. cap: per capita (per habitant)
3. e: equivalence factor
4. GHG: Greenhouse Gases
5. GJ: gigajoules
6. ha: hectare
7. gha: global hectares
8. EF: EF
9. kWh: Kilo-watt hour
10. I-O: Input-Output Analysis
11. IPCC: Intergovernmental Panel on Climate Change
12. REAP: Resources and Energy Analysis Program
13. CDW: Construction and Demolition Waste
14. MSW: Municipal Solid Waste
15. t: ton

APPENDIX C. GLOSSARY OF TERMS

1. Biocapacity: see capacity.
2. Capacity: number of people in the current circumstances on the planet could sustain over a long period of time (decades) without deteriorating the overall productivity of the land.
3. Compost: Compost (sometimes also called compost) is the product obtained from composting, and is an "intermediate" decomposition of organic matter, which is in itself a good fertilizer.
4. Direct consumption: those that generate direct consumption of resources on site.
5. Indirect consumption: those that generate indirect resource consumption, as consumption of energy or material resources come from previous resources.
6. Ecological Deficit: difference between the area available (capacity) and the area consumed (EF).
7. Sustainable development that meets the needs of the present without compromising the ability of future generations to meet their own needs.
8. Embodied energy: energy content of the materials in the processes of raw material extraction, production, processing, transportation partners, laying, maintenance and disposal.
9. Equivalence or weighting factor: provides the differences in average global productivity among different types of landscape.
10. Factor productivity or performance: compare local productivity of each category of land about a hypothetical type of territory whose biological productivity is the global average of all territories.
11. EF: tract of land that would be needed to provide the resources (cereals, fodder, fuel, fish and urban land) and absorb the emissions (CO_2) of world society.
12. Weighted footprint: EF given in standard productive hectares (gha).
13. Energy intensity: energy consumption (GJ) for each tonne of agricultural resources already available to consumers
14. Natural Productivity: amount of land (ha) required to produce 1 t of resources (food).
15. Energy Productivity: amount of land (ha) required to produce 1 GJ of energy.
16. Recycling rate: is defined as the weight of material that is recycled to the weight of waste generated.
17. Territory for CO_2 absorption: area of forest required to absorb CO_2 emissions due to consumption of fossil fuels for energy production.
18. Productive territory: see capacity.

19. Standard productive territory: average surface biological productivity worldwide. That is, the productivity factor is 1

APPENDIX D. EMBODIED ENERGY FROM DIFFFENT SOURCES

Table D1. The specific embodied energy of construction materials according to 6 different sources. Simple materials.

Material	Specific embodied energy (MJ/kg)					
Source	[18]	[6]	[20]	[17]	[16]	[13]
Commercial steel (20% recycled)	35	43	43	35-43	30.13	25
100% recycled steel (theoretical)	17					9
Stainless steel				177		
Primary aluminium	215	160	180	205	180	200
100% recycled aluminium (theoretical)	23					
85% recycled aluminium						45
Fired clay, brick and tiles	4.50				2.90	2
Fired clay, ceramics, vitrified materials	10			7.20		8
Fired clay. Sanitary ware	27.50					
Sand (aggregates)	0.10	0.10	0.10	0.10	0.08	0.50
Recycled aggregates				0.10		
Asphalt in fabric (oxy-asphalt)	10		10.00	10.00		
Lime				3.43		4.5-5.0
Plasterboard				7.90	5.73	5
Cement	7	7.20	7.20	7.00		3.6-4.0
Ceramic				2.3-2.5		
Glazed ceramic				13.00		
Primary copper	90	90	90	150		85
Fibre cement (of asbestos)	6			9.50		
Fibre cement (synthetic fibres or wood)	9			9.50		
Natural fibre				1.70		
Mineral fibre				2.35	18.40	
Synthetic fibre				30		
Fibreglass	30		30	22		35
Gravel	0.10	0.10	0.10			
Stoneware				10.90		

(Table D1) contd.....

Material	Specific embodied energy (MJ/kg)					
Source	[18]	[6]	[20]	[17]	[16]	[13]
Temperate-climate wood	3	3	3	2.10		
Tropical wood	3					
Wood, formaldehyde-free chipboard	14	14	14	14		
Wood, chipboard with formaldehyde	14					
Wood, plywood	5	5	5			
Paper				31.10		
Plastic paint (water-based) green-compliant	20					
Plastic paint (water-based)	20	20	20	20	42.23	
Synthetic paints and varnishes (enamel), organic solvent-based, organic-compliant	100			90		
Synthetic paints and varnishes (enamel) based on organic solvents	100	100	100			
Stone				0.18		0.50
Lead				190		22
Polycarbonate				79		
Polychloroprene (neoprene)	100	120	120	100-120		
Expanded polystyrene (EPS)	100	100	100	100-115		125
Extruded polystyrene (XPS) with HCFC-type blowing agent	100			100-115		133
Extruded polystyrene (XPS) with CO_2-type blowing agent	100					130
Primary polyethylene	77	75		85		110
Recycled PE (over 70%)			75			
Primary polypropylene	80		77			115
Polyurethane (PUR) with HCFC-type blowing agent or dichloromethane	70			70	82.33	135
PUR with CO_2-type blowing agent or similar	70	70	70			135
Primary PVC	80	80	80	70	53.82	85
Terrazzo				2.30		1.50
Sheet glass	19	19	19	19	16.20	
Gypsum-plaster	3.30	3.30	3.30	2.57	2.45	1

Table D2. The specific embodied energy of construction materials according to 6 different sources. Compound materials.

Material	Specific embodied energy (MJ/kg)					
Source	[18]	[6]	[20]	[17]	[16]	[13]
Hollow brick wall	2.96		2.80		2.90	
Perforated brick wall	2.85					
Solid brick wall	2.86					
H-150 concrete	0.99					
H-175 concrete	1.03					
H-200 concrete	1.10					
Prefabricated concrete				2.30		1.50
M-40/a mortar	1					
M-80/a mortar	1.34					
Prefabricated mortar				2.0-2.5	2.25	1
Aluminium windows / doors					218	
Wooden windows / doors					26.85	

APPENDIX E. MATERIALS EMBODIED ENERGY

Table E1. Construction materials embodied energy.

Construction material	u	Mm (u)	Cc (kg/u)	Cm (kg)	Eiem (MJ/kg)	Eim (MJ)
Acrylic						
Acrylic block label	u	4.0	0.27	1.0	90	96
Acrylic door label	u	218.0	0.27	58.2	90	5,238
Acrylic story label	u	28.0	0.27	7.4	90	673
Aluminium						
Aluminium folding door	m2	14.4	9.36	135.2	200	27,044
Aluminium sliding door	m2	327.6	20.00	6552.0	200	1,310,400
Aluminium grid	u	428.0	0.00	1.7	160	274
Aluminium air vent	u	667.9	0.33	220.4	160	35,266
Air conditioned input mesh	u	856.0	0.04	36.8	160	5,889
Aluminium sliding door	m2	470.2	9.36	4401.4	200	880,271
Recessed alarm button	u	12.5	0.05	0.6	160	102
Aluminium sliding door	m2	110.8	9.36	1036.8	200	207,365
Brass						
Handle brass	u	38.6	0.01	0.5	100	46
Plain key lock	u	16.2	0.22	3.6	100	356
Plain key lock	u	84.1	0.22	18.5	100	1,851
Main door lock	u	107.6	0.22	23.7	100	2,367
Brass knob set	u	107.6	0.31	33.3	100	3,334
Brass knob or handle set	u	864.6	0.35	302.6	100	30,262
Main valve diam. 8 mm	u	214.0	0.48	102.7	100	10,272
Main valve, faucet 1 1/4"	u	4.0	0.50	2.0	100	200
Main valve, diam. 1"	u	163.5	0.75	122.6	100	12,263
Main valve, diam. 3/4"	u	968.0	0.65	629.2	100	62,920
Angle main valve diam. 1/2"	u	1070.0	0.50	535.0	100	53,500
Spyhole	u	108.3	0.05	5.3	100	531
Hasp	u	38.6	0.29	11.3	100	1,134

(Table E1) contd.....

Construction material	u	Mm (u)	Cc (kg/u)	Cm (kg)	Eiem (MJ/kg)	Eim (MJ)
Cabinet brass bolt, 11 cm	u	3009.7	0.11	331.1	100	33,107
Handle	u	967.7	0.26	247.7	100	24,774
Sliding door handle	u	4.4	0.32	1.4	100	141
Door badge	u	11.5	0.05	0.6	100	58
Sliding door system	u	4.4	1.00	4.4	100	442
Brass lid of siphon sink	u	214.0	52.54	11243.6	100	1,124,356
Sluice valve diam. 1 1/2"	u	15.0	19.00	285.0	100	28,500
Ball valve diam. 2 1/2"	u	4.0	3.24	13.0	100	1,296
Sink valve with chain and lid	u	107.0	0.31	33.2	100	3,317
Main input valve of meter	u	117.0	6.00	702.0	100	70,200
Retention valve, dia. 1 1/2"	u	4.0	0.35	1.4	100	139
Meter main output valve	u	117.0	6.00	702.0	100	70,200
Cement						
White cement in sacks	t	7.1	1000.00	7110.0	7	49,770
Cement in sacks	t	173.1	1000.00	173069.0	7	1,211,483
Ceramic						
Tile 15×15 cm	u	335658.5	0.30	100697.6	8	805,580
Floor tile 14×28 cm	u	145339.5	0.74	107551.2	3	311,899
Brick 25×11.5×7 cm	mu	6.7	2100.00	14007.0	3	40,620
Brick 24×11.5×7 cm	mu	4.4	1320.00	5742.0	3	16,652
Brick 24×11.5×9 cm	mu	239.6	1550.00	371395.5	3	1,077,047
Brick 24×11.5×44 cm	mu	144.4	1948.00	281291.2	3	815,744
Brick 24×11.5×5 cm	mu	318.6	2100.00	669018.0	3	1,906,701
Perforated brick	mu	5.7	2100.00	11961.6	3	34,091
Slimed format brick	u	255.0	0.74	188.7	3	547
Plinth 14×28 cm	u	2860.0	0.74	2116.4	3	6,138
Concrete						
Block 40×20×12cm	u	22803.1	11.00	250834.2	2	501,668
Concrete ventilation duct	m	834.9	191.67	160023.4	2	320,047
Cellular concrete	m3	303.7	500.00	151870.0	1	151,870
Concrete HA- 25	m3	1271.4	2500.00	3178425.0	1	3,178,425
Concrete HA- 25	m3	2985.4	2500.00	7463462.5	1	7,463,463

(Table E1) contd.....

Construction material	u	Mm (u)	Cc (kg/u)	Cm (kg)	Eiem (MJ/kg)	Eim (MJ)
Concrete HA- 30	m3	1184.4	2500.00	2961062.5	1	2,961,063
Concrete HM- 20	m3	156.7	2500.00	391687.5	1	391,688
Concrete HM-20	m3	6.6	2300.00	15157.0	1	15,157
Reinforced concrete lid	m2	10.1	100.91	1016.2	2	2,032
Copper						
Coaxial cable TV	m	32.0	0.01	0.3	100	32
Cable 1×1 mm2/750V	m	1070.0	0.01	15.0	100	1,498
Cable 1×1.5 mm2/750 V	m	25698.7	0.02	488.3	100	4,882,760
Cable 1×16 mm2/750 V	m	10.1	0.18	1.8	100	177
Cable 1×2.5 mm2/750 V	m	43950.0	0.02	966.9	100	96,690
Cable 1×35 mm2/1000 V	m	45.8	0.43	19.8	100	1,977
Cable 1×4 mm2/750 V	m	606.0	0.05	27.3	100	2,727
Cable 1×6 mm2/750 V	m	10606.3	0.06	668.2	100	66,820
Stripped copper wire	kg	50.6	1.00	50.6	100	5,059
80 A fuse set	u	342.0	0.42	143.6	100	14,364
Triple copper wire	m	3474.6	0.02	55.6	100	5,559
50 A fuse set	u	115.0	0.06	6.3	100	633
Electric ground	u	116.0	0.10	12.1	100	1,206
Tube dia. 5/8"	m	532.3	0.33	175.7	100	17,567
Tube dia. 1 1/8"	m	562.1	0.60	337.3	100	33,726
Tube dia. 13/15 mm	m	1920.8	0.39	749.1	100	74,910
Tube dia. 16/18 mm	m	1639.1	0.48	786.8	100	78,676
Tube dia. 20/22 mm	m	2937.8	0.59	1733.3	100	173,332
Chromed tube dia. 8 mm	m	267.5	0.19	51.9	100	5,190
Glass						
Colourless glass 4 mm	m2	778.3	12.50	9728.1	18	175,106
Textured colourless 3-4 mm	m2	119.5	10.00	1194.7	18	21,505
Gypsum						
Metallic frame, 46×600 mm	m2	3835.6	2.70	10356.2	7	72,493
Metallic frame, 46×400 mm	m2	9162.4	2.70	24738.5	7	173,169
Gypsum-cardboard 13 mm	m2	21253.5	10.00	212534.5	7	1,487,742
Gypsum-cardboard 15 mm	m2	11216.7	12.00	134600.3	7	942,202

(Table E1) contd.....

Construction material	u	Mm (u)	Cc (kg/u)	Cm (kg)	Eiem (MJ/kg)	Eim (MJ)
Packaged stucco e-30	t	2.3	1000.00	2287.3	3	5,718,35
White plaster YF	t	62.6	1000.00	62599.0	3	156,497
Black plaster YG	t	187.6	1000.00	187586.6	3	468,967
Plaster for board joints	kg	10973.6	1.00	10973.6	3	27,434
Fiberglass gypsum board 13+25 mm	m2	6454.4	14.40	92943.9	7	650,608
Plain gypsum board	m2	3006.6	14.27	42912.9	3	107,282
Gypsum wall panel	m2	84.6	1.00	84.6	3	212
Lime						
Lime	t	17.8	1000.00	17841.6	4	71,366
Quicklime	t	15.1	1000.00	15130.0	4	60,520
Paint						
Synthetic varnish	kg	1503.3	1.00	1503.3	100	150,328
Surface treatment	kg	751.6	1.00	751.6	100	75,160
Solvent	l	517.8	0.86	445.3	100	44,532
Synthetic enamel	kg	527.1	1.00	527.1	100	52,709
Antioxidant primer	kg	290.2	1.00	290.2	100	29,015
Sealing gasket	m	3875.6	0.12	465.1	100	46,508
Plastic putty	kg	896.7	1.00	896.7	100	89,674
Adhesive paste	m2	7163.6	10.50	75217.6	100	7,521,759
Levelling paste	m2	20.7	10.50	216.8	100	21,683
Asphaltic paint	kg	4147.1	1.00	4147.1	20	82,941
Plastic paint	kg	733.8	1.00	733.8	20	14,676
Plasticizer	l	324.6	1.23	399.3	100	39,931
Sealer	kg	686.9	1.00	686.9	100	68,691
Mould	kg	50908.5	1.00	50908.5	20	1,018,169
Brea-epoxy treatment	m2	4782.8	3.00	14348.3	100	1,434,828
Polyethylene						
Joint 10 mm dia.	m	1685.6	0.05	84.3	85	7,164
Rope	m	535.0	0.03	17.7	85	1,501
Sheet 0.2 mm	m2	1635.3	0.19	302.5	85	25,715
Corrugated tube, d. 125 mm	m	14.7	1.56	23.0	85	1,952
Corrugated tube, d. 25 mm	m	1103.3	0.27	300.1	85	25,508

(Table E1) contd.....

Construction material	u	Mm (u)	Cc (kg/u)	Cm (kg)	Eiem (MJ/kg)	Eim (MJ)
Corrugated tube, d. 32 mm	m	2670.8	0.40	1068.3	85	90,806
Corrugated tube, d. 40 mm	m	63.7	0.50	31.9	85	2,707
Corrugated tube, d. 80 mm	m	177.1	1.00	177.1	85	15,057
Tube, d. 75 mm	m	177.1	0.94	166.1	85	14,116
Tube, d. 90 mm	m	14.9	1.13	16.7	85	1,420
Reticular tube PE, d. 25 mm	m	64.3	0.33	21.2	85	1,805
Polystyrene						
Electric cabinet, 9 elements	u	115.5	0.68	78.3	110	8,614
Telephone cabinet	u	4.0	0.50	2.0	110	220
Sockets 25A, w. ground	u	2678.0	0.06	166.0	110	18,264
Sockets 16 A, w. ground	u	107.0	0.11	11.3	110	1,248
Electric cabinet 1dif. 6 mag.	u	116.0	0.40	46.4	110	5,104
General electric cabinet	u	116.0	0.40	46.4	110	5,104
Universal electric box	u	4921.5	0.03	128.0	110	14,075
Speakers cabinet	u	36.0	0.50	18.0	110	1,980
Magneto-thermic switch	u	787.0	0.18	144.8	110	15,929
Double commuter switch	u	642.0	0.07	47.5	110	5,226
Differential switch 25 mA	u	118.0	0.23	26.7	110	2,933
Differential switch ii 40 mA	u	8.0	0.23	1.8	110	199
Simple switch	u	840.0	0.07	60.5	110	6,653
Meter cabinet	u	115.0	3.78	434.9	110	47,842
Cover plate	u	107.0	0.10	10.7	110	1,177
Interphone for 16 dwellings	u	8.0	0.65	5.2	110	572
Polystyrene plate, 12kg/m3	m3	4.7	12.00	56.3	110	6,191
Push button	u	235.0	0.05	12.0	110	1,318
Speakers support	u	4.0	1.00	4.0	110	440
Interior telephone	u	117.0	0.35	41.0	110	4,505
Porcelain						
Porcelain bidet, white	u	109.1	18.30	1997.3	28	54,925
Toilet, c. white	u	218.3	29.80	6504.7	28	178,880
Porcelain sink, white 0.50 m	u	109.1	12.00	1309.7	28	36,016

(Table E1) contd.....

Construction material	u	Mm (u)	Cc (kg/u)	Cm (kg)	Eiem (MJ/kg)	Eim (MJ)
Porcelain sink, white 0.60 m	u	109.1	12.20	1331.5	28	36,616
Porcelain pedestal, white	u	218.3	12.00	2619.4	28	72,032
Porcelain laundry sink	u	109.1	18.75	2046.4	28	56,275
PCV						
Toilet seat and lid	u	214.0	0.50	107.0	80	8,560
Outlet pipe, 110 mm	m	799.7	1.39	1111.6	80	88,931
Syphon tube, 125 mm	u	214.0	0.20	42.8	80	3,424
Syphon trap, 160 mm	u	34.0	1.98	67.4	80	5,391
Evacuation pipe	u	107.0	0.50	53.5	80	4,280
Deflector	u	107.0	3.50	374.5	80	29,960
Bath drain tube 1 1/4"	u	214.0	0.31	66.3	80	5,307
Union pipe 110 mm	m	214.0	1.49	318.9	80	25,509
Coupling 3/4"	u	117.0	0.35	41.0	80	3,276
Coupling 7/8"	u	117.0	0.54	63.2	80	5,054
Roller blind 1 mm	m2	490.9	6.07	2979.6	80	238,366
Venetian mechanism	u	236.8	7.00	1657.8	80	132,622
Flexible tube 3/4"×50 cm	m	117.0	0.05	5.9	80	468
Tube 125 mm, 4 kg/cm2	m	664.9	3.41	2263.8	80	181,105
Tube 160 mm, 4 kg/cm2	m	109.3	4.36	476.2	80	38,099
Tube 200 mm, 4 kg/cm2	m	46.0	5.45	250.7	80	20,057
Tube 250 mm, 4 kg/cm2	m	21.4	6.81	145.4	80	11,631
Tube 32 mm	m	664.6	0.87	579.6	80	46,365
Tube 40 mm	m	1053.7	1.09	1148.5	80	91,881
Tube 50 mm	m	375.0	1.36	510.8	80	40,860
Flexible tube 13 mm	m	28805.0	0.02	547.3	80	43,784
Flexible tube 16 mm	m	202.0	0.02	3.8	80	307
Flexible tube 23 mm	m	1314.6	0.02	25.0	80	1,998
Flexible tube 29 mm	m	2447.2	0.02	46.5	80	3,720
Sand						
Quartz sand	kg	4782.8	1.00	4782.8	0	383
Fine sand	m3	222.1	1500.00	333120.0	0	26,650
Coarse sand	m3	334.8	1500.00	502185.0	0	40,175

(Table E1) contd.....

Construction material	u	Mm (u)	Cc (kg/u)	Cm (kg)	Eiem (MJ/kg)	Eim (MJ)
Marble sand	kg	82555.5	1.00	82555.5	0	6,604
Gravel	m3	461.5	1700.00	784567.0	0	78,457
Steel						
Closed stirrup diam. 40 to 50 mm	u	8.0	0.05	0.4	40	16
B400S steel	kg	10.5	1.00	10.5	40	420
B500S steel	kg	234915.3	1.00	234915.3	40	9,396,613
ME B500T electro welded mesh	kg	72.9	1.00	72.9	40	2,917
Steel extruded tube	kg	7542.4	1.00	7542.4	40	301,696
S275JR, steel beam	kg	10.5	1.00	10.5	40	420
Tie up wire	kg	1078.5	1.00	1078.5	40	43,140
Steel glazed bath, white	u	109.1	25.00	2728.5	40	109,140
Pipe clamp 500 mm diam.	u	64.7	22.50	1455.1	40	58,203
Post box	u	111.0	1.23	136.9	40	5,475
Lattice for fixed slats, galv.	m2	7.7	30.30	233.3	40	9,332
Helicoidally duct diam. 500 mm	m	99.5	7.63	759.1	40	30,364
Handle for bathroom	u	3.0	0.73	2.2	40	87
Tap and fittings, bath	u	214.0	1.64	351.0	177	62,120
Tap and fittings for bidet	u	214.0	1.20	256.8	177	45,454
Tap and fittings for sink	u	107.0	1.39	148.7	177	26,325
Tap and fittings for basin	u	214.0	1.20	256.8	177	45,454
Inox. sink of 1.00 m	u	109.1	5.00	545.7	177	96,589
Shade metal frame	m	570.9	0.10	57.1	40	2,284
Flexible tubes	u	481.5	0.08	36.1	40	1,445
Stainless steel corner kit	u	214.0	0.15	32.1	177	5,682
Chromed screws kit	u	321.0	0.08	24.1	40	963
Main valve, faucet, 1 1/2"	u	2.0	0.50	1.0	40	40
Main valve 1/2"	u	214.0	0.50	107.0	40	4,280
Main corner valve 1/2"	u	214.0	0.50	107.0	40	4,280
Galv. steel frame for grid	u	856.0	0.16	137.0	40	5,478
Metallic panel 50×50 cm	u	224.7	4.63	1040.0	40	41,602
Ground copper plated, 2 m	u	16.0	1.88	30.0	40	1,202
Steel glazed shower	u	109.1	10.57	1153.6	40	46,144

(Table E1) contd.....

Construction material	u	Mm (u)	Cc (kg/u)	Cm (kg)	Eiem (MJ/kg)	Eim (MJ)
Frame for sliding door	m	4378.6	2.65	11616.4	40	464,657
Hinged steel door	m2	210.3	12.26	2577.5	40	103,100
Fire hinged steel door	m2	68.2	37.75	2574.3	40	102,971
Garage steel door	m2	20.5	39.25	804.6	40	32,185
Metallic stand 3m	u	85.1	15.70	1336.5	40	53,462
Butane gas regulator	u	107.0	12.30	1316.1	40	52,644
Iron cast grid dia. 150 mm	u	30.0	24.00	720.0	40	28,800
Steel tube 40 mm	m	24.0	4.29	103.0	40	4,118
Galvanized steel tube 1 1/2"	m	12.5	3.71	46.4	40	1,855
Galvanized steel tube 2"	m	10.0	5.22	52.2	40	2,088
Retention valve 2"	u	10.0	1.39	13.9	40	557
Stone						
Embrasure 30×5 cm	m	522.3	21.60	11280.8	0	3,384
Floor 33×33 cm	m2	7990.8	60.00	479450.4	2	958,901
Terrazzo floor 40×40	m2	8719.4	89.00	776026.6	2	1,552,053
Natural stone counter top	m	22.7	25.00	567.5	0	170
Limestone staircase step	m	586.4	20.16	11822.6	2	23,645
Limestone plinth 40×10 cm	u	1786.9	2.20	3931.2	2	7,862
Terrazzo plinth 40×7 cm	u	27883.2	1.21	33822.4	2	67,645
Limestone staircase riser	m	686.4	8.13	5580.8	2	11,162
Limestone door sill 30×3cm	m	27.6	24.48	675.6	2	1,351
Wood						
Sapelly frame 185×30 mm	m	6.4	2.72	17.4	3	52
Sapelly frame 70×40 mm	m	4457.1	1.90	8486.3	3	25,459
Sapelly frame 90×50 mm	m	527.1	3.06	1613.0	3	4,839
Standard sapelly panel 35 mm	u	680.3	11.20	7619.4	3	22,858
Standard sapelly panel 45 mm	u	108.3	11.20	1213.2	3	3,640
Sapelly panel 35 mm, with glass frame	u	223.4	3.44	768.6	3	2,306
Flanders pine lath 185×30 mm	m	6.6	1.93	12.7	3	38
Flanders pine lath 70×30 mm	m	4488.6	0.74	3299.1	3	9,897
Flanders pine lath 90×30 mm	m	542.2	0.94	509.6	3	1,529
Pinewood board	m3	34.1	425.00	14475.5	3	43,427

(Table E1) contd.....

Construction material	u	Mm (u)	Cc (kg/u)	Cm (kg)	Eiem (MJ/kg)	Eim (MJ)
Pinewood plank	m3	25.5	425.00	10850.3	3	32,551
Sapelly wood	m3	2.0	650.00	1319.5	3	3,959
Sapelly flashing 60×15 mm	m	8932.4	0.61	5448.8	3	16,346
Sapelly flashing 70×20 mm	m	1075.7	0.25	263.6	3	791
Others						
Bentonite	kg	88625.8	1.00	88625.8	2	203,839
Modified mortar	kg	150101.0	1.00	150101.0	100	15,010,096
Textile grid 12 cm	m	564.7	1.75	988.1	25	24,703
Bitumen membrane 4 mm	m2	3565.7	4.00	14262.8	10	142,628
Hard fiberglass panel 25mm	m2	213.7	1.75	374.0	25	9,349
Rubber floor,20 ×20 cm	m2	547.5	4.00	2189.8	100	218,984
Syphon bottle diam. 40 mm	u	428.0	0.45	192.6	90	17,334
Fibre cement tube 80 mm	m	11.4	4.77	54.6	10	546

SUBJECT INDEX

www.ingramcontent.com/pod-product-compliance
Lightning Source LLC
Chambersburg PA
CBHW041706210326

41598CB00007B/556